Ethics and Governance in Project Management

Small Sins Allowed and the Line of Impunity

Best Practices and Advances in Program Management Series

Series Editor
Ginger Levin

Ethics and Governance in Project Management

Small Sins Allowed and the Line of Impunity

Eduardo Victor Lopez

SKEMA Business School, Lille, France

Alicia Medina

Umea University, Sweden

CRC Press
Taylor & Francis Group
Boca Raton London New York

CRC Press is an imprint of the
Taylor & Francis Group, an **informa** business

AN AUERBACH BOOK

CRC Press
Taylor & Francis Group
6000 Broken Sound Parkway NW, Suite 300
Boca Raton, FL 33487-2742

First issued in paperback 2019

© 2016 by Taylor & Francis Group, LLC
CRC Press is an imprint of Taylor & Francis Group, an Informa business

No claim to original U.S. Government works

ISBN-13: 978-1-4987-4383-9 (hbk)
ISBN-13: 978-0-367-87489-6 (pbk)

Visit the Taylor & Francis Web site at
http://www.taylorandfrancis.com

and the CRC Press Web site at
http://www.crcpress.com

Dedication

To my lovely wife, Nandit@. E. L.

To my dearest family. A. M.

Contents

Preface

Despite all the attention devoted to governance over recent decades, project failures continue to shake the field, suggesting that further investigation is necessary. The recurrence of project governance difficulties made us question the basics of what had hitherto been taken to be unquestionable knowledge. As a result of the underlying cognitive unrest, a revision of the approaches to ethics in the project world led to the introduction of new concepts such as *Small Sins Allowed,* the *Line of Impunity,* and the *Ethics Cube.*

With this book we aim to provide business students and project practitioners (managers, team members, and other project stakeholders) with an understating of the relationship between ethics and project governance, along with the influence exerted by the context.

In order to enrich students' experience, we have included case studies with questions and reflective exercises. All the cases come from our real-life project experiences or have been collected during focus groups and workshops in the United States, Latin America, and Sweden.

Our stance is that a dissonance between business and society ethics may lead to improper behavior, which, together with outdated governance practices, may result in project failure. The ethical components of project management need to be understood and the common sources of deviance identified in order to apply new approaches to project governance that may lead to successful projects.

We start with a historical perspective of the evolution of ethics, oriented particularly toward the fundamentals of ethics that govern individual and collective behaviors within the project world. Then, aspects of the ethical approaches and contrasting views are presented and discussed. This is followed by an exploration into the fields of decision-making models, economization, double standards, honesty, and codes of ethics. Aspects of the context, such as perception, motivation, and culture, are also discussed before analyzing corporate and project governance.

At this point, the book introduces two important new concepts, the *Small Sins Allowed* and the *Line of Impunity,* which together can be the foundation for a renewed view of project governance. The *Small Sins Allowed* establish a level above which adherence to ethical standards is expected. The *Line of Impunity* relates to the perception that privileges are entitled at certain positions. Then, ethical issues and ethical dilemmas are reviewed and an "ethical hierarchy" model is introduced to explain how ethical issues should be addressed according to their moral intensity.

Later, an "Ethics Cube" is depicted to illustrate the existence of different sets of ethical values such as professional ethics, family ethics, and general ethics. It also illustrates the conflicting relationship between those sets of ethical values and personal interests, allegiances, and the presence of opportunity. This concept is expected to generate awareness about the complexity of managing multiple sets of ethical values together with other compelling factors. Shuffling the cube may result in a professional ethics affected by selfishness, idealism, pragmatism, and even by other ethical sets.

The book concludes by answering the question of how we can improve governance in projects, and presenting some guidelines for ethical leadership.

We would like to stress our disagreement with the prevalent dominant view that ethics can be steered solely by governance. We believe that ethics in project management goes beyond rules and regulations, and that people's actions have an impact on governance. We call for a paradigm shift, a new mindset understanding that professional ethics conflicts with personal factors, where rules and principles are applied selectively, and where the impact of culture and context are incorporated into the daily life of projects.

Acknowledgments

I wish to recognize the enormous support I received from Dr. Alicia Medina during the process of preparing my dissertation. I also wish to thank Dr. Ginger Levin for her confidence in my views, which contributed to the writing of this book. E. L.

My acknowledgment to all the people who contributed with anecdotes, cases, and insights, especially Thomas Larsson and the members of the PMI Chapter in south Sweden. A. M.

About the Authors

Eduardo Victor Lopez, PhD, PMP, PgMP, is Professor of Business Ethics and International Business at Belmont University (Nashville, Tennessee). He has devoted many years to studying behaviors of individuals immersed in the corporate world, from entry level to the C-suite. He got an insider view of the governance practices of many companies, including Ericsson, Siemens, Nokia, and Bridgestone. He holds a MSc in Project Management (George Washington University), and a PhD in Strategy, Program, and Project Management (SKEMA Business School, Lille, France). He works as a consultant, delivering lectures and training on project management, ethics, and governance. He can be contacted at lopeduardo@yahoo.com

Alicia Medina, PhD, is Associate Professor in Management at Umeå University in Sweden and a visiting lecturer at SKEMA in France and at the University of Montevideo in Uruguay. She is also a senior consultant and counselor in Program Management and Change Management as well as Organizational Design. Alicia has been, over more than two decades, a manager at various organizational levels in international corporations such as Ericsson, AstraZeneca, SonyEricsson, and IKEA. She holds a BSc and a MSc in Mathematics and Computer Science from Gothenburg University, Sweden, a BSc in Work Psychology from Lund University, Sweden, and a PhD in Strategy, Programme, and Project Management from SKEMA, France. She can be contacted at alicia@medina.se

Chapter 1

Antecedents

We are of different opinions at different hours, but we always may be said to be at heart on the side of truth.

– Ralph Waldo Emerson

1.1 The Beginning

When considering the source of ethics, one must take into account a number of factors, including parental examples at home, socialization in school, religious teachings, and the legal system. All of these compose a complex construct of universals (the portion of ethics norms that are widely accepted), and local beliefs (principles confined to a particular culture). Drucker (1954) saw morality manifested by real demeanor.

Relating the social values of a particular group with the ethical baggage of its individuals was a challenging task even before the concept of "rules of the game" was introduced by Milton Friedman (1970, p. 4), which made it cumbersome. He postulated: "In an ideal free market . . . [t]here are no social values, no social responsibilities in any sense other than the shared values and responsibilities of individuals." He went further by saying that "there is one and only one social responsibility of business . . . increase its profits."

Friedman's concepts emphasize the detachment between business and ethics nurtured during an age of cultural conflict that challenged the world's status quo. The 1960s (characterized by upheaval that included the Civil Rights Movement, the Vietnam War, anti-establishment hippies, and Soviet tanks putting an end to the Prague Spring) were followed by another decade of conflict, when the oil embargo, the Watergate scandal, the rise of the People's Republic of China, violence in the Middle East, and the U.S. defeat in Vietnam undermined general confidence in the existing, established values (including ethics, attachment to a culture, view of authority, and family bonds, among others).

Such periods of change, characterized by intense experimentation and high failure rates, resulted in confusion. While the public was concerned with peace and tolerance, which increased their political awareness, Wall Street ran in a different direction—straight into a fundamental difference between the mundane and the business worlds. Thus, a new attitude toward individualism arose.

During those years, many young professionals endorsed Friedman's thoughts, which were later recognized as neo-liberalism. As these professionals became corporate members and climbed the organizational ladders, the feeling of a liberal free market without an ethical commitment quickly took hold and excluded social values, leaving legality as the sole ethical bond. Much like the nonconformist Ranters in old England (around 1650), the professionals since the 1970s believed themselves to be above the moral law and felt that they were free to practice debauchery.

The 1980s adopted the language and logic of economics. Economic notions had, with their own inner logic, caused a fracture between economic reasons and moral purpose. Free-market logic subverted the meaning of morality, departing from the social and ethical values of society.

Projects as a way of organizing and performing work began during the 1950s as a need for U.S. space undertakings as well as in construction and other aerospace and defense initiatives. During the last five decades, project work has increased continuously, and methods, tools, and processes have been developed to implement it. Several disciplines and knowledge areas have been established within project management. There are a large number of tools for project planning, staffing, risk management, stakeholder management, and communication; however, there is not a single tool that provides support regarding ethical issues. The project management associations that establish standards (including the Project Management Institute [PMI], the International Project Management Association [IPMA], Axelos [PRINCE2], the Global Alliance for Project Performance Standards [GAPPS], the Australian National Competency Standards for Project Management [ANCSPM], the Australian Institute of Project Management Standards [AIPM], the South African Qualification Authority [SAQA NQF Level 5], and the Project Management Association of Japan [P2M]) focus on matters related to project management tools and processes, with no mention of project ethics beyond the codes of ethics. Dr. Thomas Grisham (2011, p. 1) commented that "neither the PMI, nor the IPMA, deal with the issue of bribes, which are common in many countries, and present in most. On international projects how is a project manager to deal with a local official who will not clear equipment through customs without a 'tip'?"

1.2 Contrasting Views (1): "[People] can do good but only at their own expense."

When Friedman (1970, p. 3) asserted that "[people] can do good, but only at their own expense," he was sharing "Adam Smith's skepticism about the benefits that could be expected from those who affected to trade for the public good."

This first view is shared by many managers. They believe that in order to win the respect of others, it is necessary to have a moral imperative through ethical content embedded into corporate visions; in this case, ethics is traded to obtain respect. Others proposed that a company could benefit from adopting an ethical approach to management, and that competitive advantage leading to increased returns could come from the creation of this kind of intangible assets. This view suggests the notion that ethics has a trading value related to financial proceeds.

There were those who proposed that demonstrating ethical and socially responsible behavior might lead to the necessary impetus for change, although they also recognized that financial performance is almost the exclusive parameter for measuring business success.

Drucker (1986, p. 256) stated: "There are important areas where managers, and especially business managers, still do not realize that in order to be permitted to remain autonomous and private they have to impose on themselves the responsibility of the professional ethic. They still have to learn that it is their job to scrutinize their deeds, words, and behavior to make sure that they do not knowingly do harm."

After Friedman's postulate, Drucker (1993) proposed a different approach considering that this view could alienate workers and affect their motivation, but his point of view was still concerned with efficiency rather than ethics, representing the typical thought of the 1990s, when ethics at face value was associated with its capacity to improve the bottom line.

Most of the publications related to strategic planning throughout the 1980s and early 1990s omitted any mention of ethical and moral terms. Any advice to build strategy on the basis of ethical reasoning was dismissed; instead, strategy and ethics were considered separate and unrelated matters.

Most of the Charters of Fundamental Rights and Freedoms across the globe recognize that "everybody may do whatever is not prohibited by law and no one may be forced to do what the law does not command" (Maddex, 1995, p. 249). This constitutes a necessary minimal approach to set up a standard across borders, avoiding engagement in moral controversies that are due to cultural differences.

What differentiates this attempt at establishing a global minimal standard with the ethical disengagement triggered in the 1970s is that while the Charters of Fundamental Rights and Freedoms offer some minimal standards with the intention to improve the status quo, some corporations chose to adhere to the minimal standard defined by law with the intention to decrease their exposure.

In an ever-changing world, in which technology accelerates the spread of socially defined values, the law is condemned to be a laggard, as lawmakers are not always able to keep pace with the changes. For this reason, attachment to the law as the exclusive ethical justification is not enough. Legal behaviors alone may not be generally recognized as ethical. There is a disconnect between the spirit of the law (morality) and the letter of the law (legality). As legality could not take into account every perception of morality, an ethical decision is one that is legal and morally satisfactory.

During the 1980s it was observed that there was increased attention to organizational ethics, although ethical decision making in organizations was still underrepresented.

A study by Robertson (2008) of the number of ethics-related articles that appeared in the *Strategic Management Journal* between 1996 and 2005 showed that immediately after the scandals surrounding Enron and WorldCom there was increased interest in the topic, although it lost momentum soon thereafter. The study also showed that the link between governance and ethics was underrepresented.

Different approaches, even disconnection between them, were the logical result of the state of confusion previously mentioned. One iconic lyric from 1969 said: "Confusion will be my epitaph, as I crawl a cracked and broken path. If we make it we can all sit back and laugh, but I fear tomorrow I'll be crying."*

1.3 The Unsatisfactory Equilibrium

The expression *unsatisfactory equilibrium* was coined by Elkington and Hartigan (2008, p. 6) and refers to palliative endeavors, resulting from confusion, that are unable to shape the future in the social entrepreneurship field. It addressed the confrontation between those who embraced a new set of public values and those who tried to disguise their business with them. The authors claimed that a transformation is possible only when society needs are met.

---

* Epitaph (lyrics) Album: *In the Court of the Crimson King,* by King Crimson (1969), E. G. Music, Inc.

There are two reasons for business to engage in ethical practices: a desire to do the right thing and a desire to convince stakeholders of the ability to do the right thing. In the first case there is no external pressure or government constraint; the ethical motivation comes from universal and cultural moral values. In the latter case the motivation is to mitigate the risk of legal consequences.

The need to connect ethics with management practice became evident, as supporting ethical principles should not contend with having a competitive posture. Shaping a firm exclusively in financial terms contributes to the subordination of ethical concerns. Management should take into account more than just profit maximization and go beyond the short-term bottom line by looking into transcendent and everlasting values.

Public consciousness, attitudes, and opinions have evolved to a point at which, in many instances, public values differ from those of industry. As a consequence, achieving project objectives requires an alignment with the prevailing public belief, which may differ in different countries. The generalization of sustainability reports, through which companies show their commitment to transparency in times of stakeholder scrutiny, is an example of this alignment. They shed light on a new era in which public opinion becomes a key stakeholder in its own right and is worth greater attention. As a counter-example, it can be mentioned how the Botnia project (Uruguay) confronted fierce opposition from the Argentine stakeholders, generating tensions between the two countries that were only eventually resolved at the International Court of Justice (see Case 1-1).

During the 1990s, publicly regarded values included self-fulfillment, fairness, cultural identity, and environmental considerations. In 2015 there was a general acknowledgment that morality is socially constructed by the participation of all members of a community. As such, the integrity of common purpose should be an integral component of the project planning process.

Case 1-1 Botnia Paper Mills and Prevailing Public Beliefs

Despite objections related to pollution by the Argentine government and the local residents of the city of Gualeguaychu (Argentina), Botnia paper mills were built in Uruguay, alongside the transboundary Uruguay River, in 2006.

While Argentina brought the issue to the international arena hoping that further research would support its case at the International Court of Justice, protesters demanded either the suspension of Botnia's activities or their relocation farther away from Argentine territory.

A process of information exchange and dialogue among the various stakeholders was conducted by Green Cross International. One of its objectives was to enlighten them about the potential environmental impact of the paper mills. In order to do so, Green Cross International conducted technical, social, and environmental assessment studies that were independent from the governments and other stakeholders. The technical research tried to unveil the influence of the paper mills on the quality of the air, soil, and water in the region, which could potentially affect biodiversity, tourism, and the quality of life of its residents.

Although the first studies showed that air quality following the paper mills' establishment did not reflect any significant pollution levels, opponents of the project expressed dissatisfaction with the results, as many Gualeguaychu residents feared that contamination levels were exceedingly high and could affect the local tourism industry.

Later developments in this case included the blockage of an international bridge between the two countries by the local Gualeguaychu population, a refusal from the Argentine government to the use of force to clear the bridge, and the subsequent disruption of the international flow of goods for about two years. Eventually the International Court of Justice ruled in favor of Uruguay, and the bridge was cleared.

This case demonstrates how much public consciousness has evolved from the times when the industry did not care about the quality of life (as during the Industrial Revolution in England) to the present day, when voices are raised defying unconsented decisions. The evolution of attitudes and opinions had reached a point that, when public values differ from those of the industry, a confrontation is inevitable. As a consequence, an alignment with the prevailing public beliefs is imperative.

Questions

1. Do you believe that the results of technical studies are enough to appease fear?
2. Is it smart to expedite a project by complying exclusively with the law, regardless of stakeholders' expectations? Is it ethical?

1.4 Contrasting Views (2): "Conforming to social values is the moral imperative that legitimizes entrepreneurship."

Today, it seems as if Friedman's assertion is no longer generally accepted as it was in 1970. Instead, conforming to social values is the moral imperative that legitimizes entrepreneurship. It is now recognized that social expectations may not be exclusively related to legal compliance, and that meeting society's expectations for conscientious and proper behavior is the project's ethical responsibility.

A historical perspective paralleling the views of business ethics and values is presented in Table 1-1. It consists of an update based on Joyner and Payne (2002, p. 302).

A convergence between legal and moral liabilities is becoming real today as the existing "compliance with the law" driver gradually gives space to an emergent "social accountability" driver. Older concepts, such as the court of law, letter of the law, and money, are shifting to the court of the public opinion, spirit of the law, and goodwill.

Society, organizations, and projects should not necessarily pursue different goals, as what is good for one does not need to be bad for the other and vice versa. Some forms of social issue participation, as well as improving working conditions, are now recognized as international standards.

When an ethical component is included in the vision of a project (as an appeal to the concepts of social good), it can move stakeholders at large to support this vision for reasons beyond those arising only from a financial nexus. The ability to fulfill the project's social purpose influences its success.

Therefore, Jackson and Nelson (2004, p. 2) sketched out the strategies for building shareholder and societal value as concentric dimensions expanding from a primitive core, which implies compliance only by obeying the law. A second dimension includes controlling costs, risks, and liabilities. These two internal dimensions are exclusively related to former shareholder value creation. A third dimension of sustainable value and long-term growth is related to community investment. Then a fourth dimension is related to value creation that meets societal needs. These two middle dimensions, the third and fourth, relate to the feeling that there is something else to be concerned with beyond the bottom line; nevertheless, they are still attached to the expected financial returns. Finally, there is an external dimension of collaboration needed in order to solve complex social and environmental issues, which would indicate a paradigm shift from financial return to social involvement (see Case 1-2), meaning that one way of looking at the world is replaced relatively quickly by another that is due to a change in perception accompanied by a change in values.

Table 1-1 Historical Perspective of Business Ethics and Values

Author	Ethics	Values
Barnard (1938)*	Morals are active results of accumulated influences on persons evident in actions.	Responsibility: Power of private code of morals to control individual conduct.
Simon (1945)*	Ethical propositions assert "oughts" rather than facts.	Firm survival involves adapting objectives to values of customers.
Drucker (1954)*	Morality must be principle of action exhibited through tangible behavior.	First responsibility to society is to make a profit.
Selznick (1957)*	Definition of mission includes wider moral objectives.	Leadership requires defense of critical values.
Andrews (1971)*	Firm is defined only in financial terms, leading to subordination of ethical concerns.	Ethical behavior is a product of values.
Freeman (1984)*	Concern for ethics is necessary, but not sufficient to decide "what we stand for."	Enterprise strategy: What we stand for?
Jones (1991)	*Moral intensity . . . captures the extent of issue-related moral imperative in a situation (p. 372).*	*Not all people see themselves as independent moral agents in work situations (p. 390).*
Joyner and Payne (2002)	*Ethical and socially responsible behavior can boost financial results (p. 310).*	*Ethics, values, integrity, and responsibility are required in the modern workplace (p. 297).*
Anderson and Smith (2007)	*The social constructs of public perception entails examining both moral means and moral ends (p. 479).*	*Context . . . determines what values are accepted. Each society can be expected to have different values (p. 494).*
Kvalnes (2014)	*The circumstance approach to honesty focuses less on the individual decision maker's moral beliefs and convictions, and more on the environment in which the decisions take place (p. 593).*	Honesty in projects.

* Extracted from Joyner, B. E. and Payne, D. (2002). Evolution and Implementation: A Study of Values, Business Ethics and Corporate Social Responsibility. *Journal of Business Ethics* 41, 302 (Table 1).

Case 1-2 The Water Project

The Water Project (www.thewaterproject.org) is passionate about unlocking people's potential and is convinced that providing access to safe and clean water not only improves people's quality of life but also increases their chances of development. This organization helps communities around the world that lack access to clean water and good sanitation by ensuring that they are directly involved in projects that provide them with safe water.

While a community's needs are always considered first, The Water Project works with local teams and partners who have long-term relationships and commitments with the people they serve to develop clean water programs, including wells and rainwater catchment.

A paradigm shift from financial return to social involvement is perceived all around the world, from volunteer work to charities.

Taking this into consideration, please provide examples of the following:

1. Initiatives to solve social issues
2. Initiatives to solve environmental issues
3. Changes in perception accompanied by changes in values
4. Your own social involvement initiatives

The public mindset changed from the old-fashioned shareholder value creation by going beyond the bottom line and moving toward a paradigm shift. Table 1-2 shows the parallels among dimensions, actions, and the predominant mindset along this evolution.

Table 1-2 Mindset Evolution

Focus on	Action	Mindset
Compliance	Obey the law	Old-fashioned shareholder value creation
Control	Cost, risks, liabilities	
Community investment	Strategic philanthropy	Going beyond the bottom line
New value creation	New products, services, processes, business models, markets, alliances	
Collaboration	Social and environmental issues	**Current paradigm shift**

Table 1-3 Evolution of Business Ethics and Its Influences

Year	Past	1970	1980	1990	2000	2010
Era	Wonderland	Confusion	Confusion Unsatisfactory equilibrium	Unsatisfactory equilibrium	Scandals	Re-foundation
Strategy School	Design	Planning Positioning	Positioning/descriptives	Positioning descriptives	Descriptives	Learning?
World Facts	Corporate jobs Television Rock 'n Roll	Watergate Vietnam Oil embargo	Easy money Glasnost AIDS	Internet	Enron WorldCom AIG Lehman Brothers	Alternative energy High debt
Result	Optimism Individualism	Confidence in old values undermined	Transition to new values	Optimism Individualism	Recession Government intervention	Innovation Entrepreneurship
Ethical Driver	Social responsibility	Rules of the game	Free-market logic	Ethic's face value	Global concern	Social and global responsibility
Ethical Behavior	Fair play	Disengagement between business strategy and ethics/social responsibility	Divorce between Main Street and Wall Street	Dress up as committed with ethics to get competitive advantage	Public involvement	Taking a side Advocating
Key References	Barnard (1938) Simon (1945) Drucker (1954) Selznick (1957)	Friedman (1970) Andrews (1971)	Lorange (1980) Quinn (1980) Porter (1980, 1985) Ohmae (1982) Miles (1982) Freeman (1984) McCoy (1985) Hamermesh (1986) Prahalad and Doz (1987) Burgelman and Maidique (1988) Freeman and Gilbert (1988)	Brady (1990) Rumelt et al. (1994) Ghemawat (1991) Jones (1991) Mintzberg and Quinn (1991) Drucker (1993) Kennedy (1993) James (1994) Hosmer (1994)	Kelemen and Peltonen (2001) Hillman and Keim (2001) Joyner and Payne (2002) Stevens et al. (2005) Anderson and Smith (2007) Elkington and Hartigan (2008) Robertson (2008) KPMG (2008)	Kvalnes (2014)

1.5 Summary

As the world of ideas evolves, views of the way projects should address ethics progress from confining ethics to the minimal law requirements, through hybrid unsatisfactory arrangements meant to reduce legal exposure or to show adherence to ethical standards not fully embraced, to a contemporary inclusion of social values within the project's purpose (see Table 1-3).

The trading value attached to ethics, mainly related to financial returns, was manifested when many questioned: Is there any value in ethics for stakeholders (mostly for shareholders)? Departures from this view considered ethics as a motivational factor for employees, whereas others pondered the benefits of compliance to ethics and strict adherence to the minimal standards defined by law.

As public and industry values differed, debate and confusion led to an unsatisfactory equilibrium between the embracing of a new set of public standards and the pretense of doing so. On one side of the scale were those who believed in doing the right thing for its own sake; on the other side were those who pretended to do the right thing but lacked conviction about its fundamentals.

The old concept of doing good at your own expense developed toward conforming to the social values as an imperative that legitimizes businesses. Nevertheless, even though it is true that signs of a paradigm shift can be seen everywhere, it is also true that scandals did not retreat. It appears that the times of unsatisfactory equilibrium are still here, and any perception of evolution is still nothing but a dream.

A deeper understanding of this dynamic requires an exploration of three key concepts: ethics, context, and governance, which are covered in the next three chapters. In Chapter 5, two new concepts will be introduced: the Small Sins Allowed and the Line of Impunity. The application of these concepts to ethical issues and ethical dilemmas will follow in Chapter 6. After an examination of this dynamic, Chapter 7 covers the ethical cube, and finally, Chapter 8 provides a review summary and closing thoughts.

Chapter 2

Ethics

It would be as foolish to expect that our moral and ethical systems would turn out virtuous, noble, and holy beings, as that our esthetic systems would produce poets, painters, and musicians.

– Schopenhauer

The terms *morality* and *ethics* have often been used interchangeably as behaviors coherent with principles that set what is good or bad.

Morality (from the Latin *moralitas*) differentiates between right or wrong intentions, decisions, and actions. It might come in the form of a body of principles (philosophical, religious, and cultural) or standards believed to be general.

Ethics (from the Greek *ethikos*) relates to "custom and habit"; it involves systematizing and recommending what kinds of actions are right or wrong in specific contexts. Ethics is described as a moral obligation and behavior that guides decisions and actions. It also considers: the meaning of moral propositions and how their values may be determined, the practical means of determining a moral course of action, and what a person is obligated to do within some particular domain (such as a project), or in some specific situation.

Ethical virtues are related to the superior "end" of improving the lives, well-being, and happiness of individuals and society at large. *Metropolis,* the classic 1927 Fritz Lang film, depicted metaphorically that hands and head need to be joined by the heart. Ethics is "good action" oriented, and so it is the heart that joins practical and theoretical virtues.

Hence, ethics is defined as the right and just (moral) conduct or behavior of individuals and groups. As such, ethics is concerned with the judgments demanded in moral decisions which entail that something is good or bad, right or wrong.

> *Morals are the personal values and behaviors of individuals. . . . Ethics are the more systematic categorization of morals, the socialized moral norms that reflect the social systems in which morals are embedded* (Anderson & Smith, 2007, p. 480).

Knowing what is ethical does not mean doing what is ethical. We base our decisions on our likes and dislikes. Emotions, purpose, ego strength, and courage to overcome negative feedback come into play in such a way that moral judgment alone is not sufficient condition for moral action.

The concepts of integrity (honesty, sincerity, and candor), justice (impartiality, sound reason, correctness, conscientiousness, and good faith), competence (capable, reliable, and duly qualified), and utility (quality of being useful and, philosophically, providing the greatest good for the greatest number) are fundamental ethical principles (Raiborn & Payne, 1990).

> *While idealists believe that most people do seek good and right choices, and most people agree on similar definitions of what constitutes ethics, morality, and virtue, skeptics talk about a special ethical outlook for the business game, departing from the strict truth, into half-truth and misleading omissions in order to overcome obstacles* (Lopez, 2015, p. 35).

Regarding individual and group ethics, "real moral dilemmas are ambiguous. It is fair to ask if there is a collective ethic outside the individual's ethics" (Lopez, 2015, p. 35). A project's ethics can only be strong when its members' shared ethics are also strong, as illustrated in Case 2-1.

Case 2-1 Apple's Ethics

One of the pillars on which Apple bases its success is "demonstrating integrity in every business interaction." They believe that the four main principles that contribute to integrity are honesty, respect, confidentiality, and compliance.

Apple's code of business conduct applies to all its operations, including those overseas, as they want to ensure that both employees and business partners display appropriate conduct in all situations. Additionally, policies and guidelines related to corporate governance, conflict of interest, and how to handle questionable conducts were also made available, as well as a Business Conduct Helpline used to report misconduct to the Audit and Finance Committee.

As many of Apple's products and components are manufactured by suppliers located in countries with low labor costs, potential for misconduct exists due to different labor and oversight standards. In order to cope with this possibility, each supplier must sign and adhere to Apple's "Supplier Code of Conduct." Suppliers' audits are mandatory to ensure compliance. No supplier can do business with Apple without abiding by its standards. Annually, Apple releases a "Supplier Responsibility Report" explaining its supplier expectations as well as the conclusions of its audits and the corrective actions taken.

Nevertheless, from time to time claims of unfair working conditions related to Apple, Nike, and other companies with a presence in low-labor-cost countries appear in the press.

Questions

1. Do you believe that companies like Apple or Nike are doing anything unethical by operating in countries with low labor costs or lax labor standards?
2. Do you think that these companies are doing all they can to ensure fair and ethical practices all across the globe?
3. What else could they do?
4. Whose ethics should apply in cases of different labor standards, such as child labor or unfair labor practices, the home country or the host country standards?

As personal moral standards are neither legislated nor can be changed by decree, there are a number of behaviors that make us angry but that are not necessarily illegal.

Ethics is seen as a definition of right and wrong, and an arrangement of value practices and principles. Laws instead are conceived to reflect society's positions and desires for the culture in which it exists. The legality (letter of the law) of an issue does not always reflect its perceived morality (spirit of the law). Ideally, each member of society should stress moral actions regardless of the law, as no matter how deep and wide a legal code is, not every immoral

> *or illegal behavior can be proscribed. Laws and legal environments make, restrain, determine, alter, delineate, and authorize organizations, providing forums and guidelines for interaction* (Lopez, 2015, p. 101).

Sometimes ethical and legal responsibilities are clearly differentiated, with areas that belong exclusively to the legal (laws and regulations) or to the ethical (philanthropy, community investment) fields; still, it is also common to find an overlap between them (product safety, environmental protection, employee health and safety, honoring contracts, fair competition, responsible marketing, and nondiscrimination).

There are two major types of ethical breakdowns: "conscious transgressions (the individuals choose to follow the unethical path) and unconscious transgressions (the individuals do not even realize that they are making an inappropriate decision, as they fall prey to ethical fading or to other cognitive biases)" (Lopez, 2015, p. 36). See Case 2-2.

Case 2-2 A Wedding and the Long Distance Calls

Back in 1999 in Buenos Aires, a group of optimization engineers working for a local cellular telecommunications provider (Miniphone) got unlocked phones to conduct their daily tests on the network. As the work to be conducted involved only local calls, none of the engineers was supposed to make any long-distance calls, and this was the case until . . .

One day Ms. Gay, who was well known for her solid stand on moral and social issues of all kind, and who several times confronted co-workers about their not-so-strong points of view on such matters, brought the good news. She planned to get married in three months. Then the wedding provisions formally started.

Although both the bride and the groom lived in Buenos Aires, their families, friends, and acquaintances resided in two provinces far from there.

One or two months later, the optimization manager was struck when his cost center was charged a stratospheric bill for long-distance calls originated from her device.

Did Ms. Gay feel entitled to break the trust deposited in her, because of her particular situation? Did she believe nobody would notice? Or did she simply rationalize that she was doing nothing wrong?

Ms. Gay's other ethical values were not affected, as she was the prey of "ethical fading" related exclusively to her work phone usage. She was not doing this using other people's phones reachable outside of her work, which could have compromised her personal set of values. She only compromised the work-related set of values. Outside of her job she continued to be a 100 percent honest person.

All Ms. Gay's co-workers attended the wedding, as she was (and still is) a very much beloved friend, although nobody else has had long-distance privileges since then.

Questions

1. Have you ever witnessed a similar situation? Explain.
2. Is there a problem with holding different ethical standards in the workplace versus in one's personal life?
3. What about when one is operating in a country with different ethical standards. Does one follow one's own ethical standards or adapt to the local situation?

In relation to conscious transgressions there are two positions. People such as Lee (1990, p. 20) believe that "individuals and groups are opportunistic and may seek to break agreement to further their own ends." In contrast, Harris and Bromiley (2007, p. 352) said that "people may prefer honesty, but can be tempted," proposing that the magnitude of the temptation influences whether a frail individual buckles under enticement and cheats. They believe that it is easier to be ethical when the incentive is minor and harder when a substantial incentive is at stake, remarking that even accepting that managers in general act ethically, the likelihood of managerial impropriety increases with the intensity of the motivation (Lopez, 2015, p. 68). This correlated with Ms. Watkins' words: "Pay and bonuses were above market, including clerical staff. If Enron paid peanuts, I doubt the fraud would have happened because people focused on their compensation, their big bonuses, getting the stock price up" (Beenen & Pinto, 2009, p. 277).

In relation to unconscious transgressions, we must consider that people make decisions that are frequently decoupled from rationality, even when maliciousness, stupidity, or neglect are not the drivers of their behavior. Whether these unconscious moral transgressions are isolated or systematic could be a matter of discussion. Ariely (2008, p. xxx) mentioned that "irrational behaviors of ours are neither random nor senseless. They are systematic and . . . predictable." By the same token, these unconscious moral transgressions follow a pattern dictated by individual moral biases, and by doing so they are predictable.

> *Increased research efforts in the field of unethical and/or illegal behaviors were triggered by the Enron, WorldCom, Tyco, HealthSouth, and other scandals. They were attributed to two main causes: organizational characteristics and employee malfeasance, pointing to attributes such as bureaucracy, culture, leadership inadequacies, financial incentives, job stress, social networks, and several economic factors, as affecting the propensity of wrongdoing* (Lopez, 2015, p. 37).

In the majority of these ethical wrongdoing situations, the inability to discern ethical behavior from outward appearances was evident, as it was difficult to distinguish who was ethical and who was not.

Ethics must be cardinal to the overall management of any project, instead of merely a peripheral subject. Substantial widespread interest in ethics by the general public is demonstrated by the fierce outcry opposing the almost daily sequence of scandals.

Some basic principles that could form the cornerstone of project ethics include honesty, good faith, transparency, and averting conflicts of interest. The Project Management Institute (PMI) "Code of Ethics and Professional Conduct" mentions honesty, responsibility, respect, and fairness.

Leadership depends on ethical behaviors, as they reduce risk exposure, boost trust, and improve positive results and long-term success. "In past years, secrecy was regarded by bankers as a virtue and puzzlement as a respected technique for controlling financial markets. Increasingly, public demands for increased accountability have led to one of the most significant changes in central banking in recent decades—the understanding of the merits of transparency" (Lopez, 2015, p. 38). Today, speaking clearly is more effective than muttering.

"When thinking about unethical situations we commonly envision large-scale incidents such as the Ponzi schemes, mortgage fraud cases, or security exchange violations, but unethical acts may occur at all levels" (Lopez, 2015, p. 38). Indeed, projects do not need to be large or complex to embrace ethical standards. For example, a small project to evaluate a few companies in a specific area to recommend for inclusion in a company's supplier list may embrace ethical issues if someone in the project has personal interest in one or more of those companies, if there is a previous relationship with one or more of these companies and the project's evaluation is based on that relationship, or if we give an evaluation based on something else than an unbiased set of criteria.

"Individuals have minimal discretion when they are faced with ethical situations that are due to the existence of tremendous pressures to meet project expectations and to adhere to institutional procedures and cultural or organizational norms" (Lopez, 2015, p. 38).

The complexity of ethical matters derives from the many biases (personal, organizational, societal, or cultural) that can influence a decision. "In numerous cases, it is easier to make unethical choices because of the simultaneous interaction of a combination of factors such as competition, culture, executive values, and opportunity" (Lopez, 2015, p. 39). "Everyone else does it" is a rationalization for making a poor ethical choice by herding (assuming that something is good based on other people's behavior). This process is called *differential association*, and it describes how the "exposure to postures favorable to vicious acts leads to the infringement of rules" (Schaefer, 2005, p. 183). Although acceptable conduct can be ruled, it could be challenged by friends or fellow employees, who may promote rather dissimilar rules. Social control and other concerns may help prevent deviant behavior in society. It is challenged by the fact that people frequently experience contending messages about how to act (see Case 2-3).

Case 2-3 Disposal of Surplus

Are ethics and ethical issues dependent on the industry? Is it possible to have different views of the same action just because we tend to apply different lenses to the same phenomenon, depending on the industry?

When we met Catharina at a workshop organized by PMI Sweden to discuss ethics in projects, she shared with us her experiences and her conclusions about the above questions.

Catharina worked for many years in the construction industry and managed projects of different magnitudes—sometimes quite small projects that involved only one building, some other times large projects aiming to deliver a whole new community. The companies were also of different sizes and operated in different modes; some were small regional companies while others were part of national or even global corporations.

In spite of all those differences, there was something that seemed to be a common norm: *to order and buy more material than it was estimated was needed.* How this extra material was used depended on the situation or the company. The surplus after the projects were completed was declared "waste" and never returned, or it was resold, or, rarely, transferred to another project. Normally, the material ended up in the hands of the workers, who were allowed to take it home. How could this be conciliated with the policies related to robbery?

This practice also affected the fair pricing of different projects, as sometimes, when "waste" from one project was available for the next project, the cost estimations that resulted were deflated, gaining competitive advantage for the latter project at the expense of the first one.

Catharina also worked as a project manager at Systembolaget, a company owned by the Swedish government that has a monopoly on the sale of liquors, wine, and all other alcoholic beverages in the country. Due to the strict views of alcohol trading in Sweden, the company has a zero-tolerance policy regarding waste—it has to be reported, investigated, and documented. All the employees were very much aware of the policy and they acted in consequence, so that waste was a rare event there.

So why is the same phenomenon considered in different ways? According to Catharina, it is because the "rules of the game" are different in different industries, and this somehow shapes people's attitude. It seems that people tend to act according to the common norms or views within their industry. These views may differ greatly from one industry to another.

Questions

1. Do you believe that when managing a project it is fair to buy material in excess just to feel safe? Could you manage the project in any other way? Explain.
2. What should you do with the surplus in such situations?
3. Should you allow workers to take excess materials declared "waste" home as a means of rewarding them for their hard work on a completed project? Explain.
4. Should the same set of ethics apply to all industries, or is it possible for ethics to vary by industry? Explain.

When the mechanisms of corruption are spread across a project, unethical behavior can subtly grab anyone, entrapping even morally solid people in a network of corruption, as in the case of Enron, where higher-than-market salaries and bonuses turned people tolerant of lax accounting principles.

Many project managers suffer "managerial myopia," as a result of which they are willing to sacrifice long-term value for short-term rewards (Harford et al., 2015). The best predictor of ethics conformity is the strength of one's commitments and belief in the moral order. Project ethics must include the principles and standards that guide conduct in the workplace.

Projects may be ones in which dis-benefits result, such as pollution, health and safety violations, accounting fraud, bribery, and corruption. "Although legitimacy originally meant being in accordance with the law, current usage has expanded its meaning to refer to the wide set of values that provide moral support" (Lopez, 2015, p. 39). Thus, it is now generally perceived that actions are proper and desirable among some socially fabricated system of norms, values, and beliefs. Hence congruence between actions and the socially set standard of behavior is required for those actions to be recognized as legitimate (Anderson & Smith, 2007).

Regarding projects and the law, there are two (meta-theoretical) perspectives. The *rational materialist* perspective sees projects as wealth maximizers and the law as a system of incentives and penalties. The *cultural normative* perspective sees projects as cultural rule followers and the law as a system of moral principles, roles, and symbols.

"The difference between legal (letter) and moral (spirit) obligations, which at some point overlap in a gray zone, is that while the former are existing (legal) and compliance driven, the latter are emerging (moral) and accountability driven" (Lopez, 2015, p. 40). Legal obligations include honoring contracts, paying taxes, complying with the law, and adhering to codes and regulations. Moral obligations go beyond them in that they add a sense of humanism not always contemplated in the legislation, as is the case of care for the environment, sustainability, inclusiveness, and solidarity.

2.1 Ethical Decision-Making Models

While the main ethical thoughts of the last decades were reviewed in the previous chapter, and the importance of project ethics was discussed, there were few references about the ethical decision-making models, which play a key role in understanding of ethical behaviors.

The ethical decision-making models (Dubinsky and Loken, 1989; Ferrell and Gresham, 1985; Hunt and Vittell, 1986, Jones, 1991; Rest, 1986; Trevino, 1986) explore different avenues attempting to shape the ethical decision-making process by recognizing the characteristics of issues that influence the process (Lopez, 2015, p. 43). Most of them share some common characteristics. These models recognize that the context (organizational, cultural, economic, and social) influences the recognition of a moral issue, which is the first milestone in ethical decision making. Only after the moral issue is recognized can a moral judgment be made. Next, with the inclusion of factors such as opportunity and individual moderators, the models advance to the establishment of a moral intent to finally engage in a moral behavior (ethics).

These "ethical decision-making models do not include intrinsic characteristics of the moral issue itself; instead they treat the moral behavior and decision-making processes of individuals in a general and unspecific way, independent of the moral issue" (Lopez, 2015, p. 41). It was Jones (1991) who recognized that the ethical decision-making process as an issue-contingent matter, meaning that the characteristics of the moral issue (*moral intensity*) are important determinants in this process. Evidence that "different types of moral issues correspond to different modes of moral reasoning" supports this line of reasoning. "As the issue changes, the consensus concerning appropriate ethical conduct is likely to change as well" (Lopez, 2015, p. 41).

> *Moral intensity is expected to vary from one issue to another. It accounts for the degree of moral issue-related factors, including the magnitude of consequences, probability of effect, social consensus, proximity, concentration of effect, and temporal immediacy; it does not contain either traits of the moral decision maker (agent) nor project factors, such as moral development, ego strength, field dependence, locus of control, values, culture, or policies"* (Lopez, 2015, p. 42).

A contemporary view regards the public at large as a participant stakeholder in project ethics. As such, the public demand for accountability requires that lapses in project governance be considered as ethical lapses.

2.2 Economization

The cornerstone of the economization of ethical thinking can be found in Friedman's words (1970, p. SM17): "there is one and only one social responsibility of business—to use its resources and engage in activities designed to increase its profits so long as it stays within the rules of the game, which is to say, engages in open and free competition without deception or fraud." Many people believe that this "free market logic [has] undermined the social and ethical values of society" (Anderson & Smith, 2007, p. 479). "However, it is simplistic to believe that any other system has incorruptible social and ethical society values. History shows how many civilizations perished under the burden of their excesses" (Lopez, 2015, p. 44).

"The crisis in business ethics has been explained from different perspectives, including management education lacking moral grounding, an organization-centered understanding of business in detriment of a more humanistic view, leadership failure, failure to recognize individuals as the unit of moral analysis, and an organizational-level corruption" (Lopez, 2015, p. 44). Huehn (2008, p. 823) said that "the economization of management has a detrimental effect on the practice of management and on society at large." He added, "economism acts like a corrosive destroying not only the basis of good (ethical and effective) management practice and theory but also the social context in which management exist."

Not all stakeholders are equal in the eyes of the project manager. One way to classify stakeholders into two distinct groups is to recognize them as market and nonmarket. Market stakeholders interact with the project through some form of economic transaction and play an integral role in the project's value chain. Nonmarket stakeholders are those who interact with the project on a noneconomic basis.

> *Many industries (commercial and financial banking among others) encounter banality, frivolity, and superficiality, in their projects. A culture of showing wealth and power is generally accepted across the board. From dress codes to job titles, a demonstration of status reigns over other values. Operators compare their shoes, watches, cars, and suits brands, amid their references to refined tastes for expensive wines, paid by the project's budget. Lavish Las Vegas conventions and discretionary expending are common, while austerity is seen as a signal of weakness"* (Lopez, 2015, p. 46).

Some describe such projects as alliances where everybody engages in his own interest. When people are willing to forfeit a bit of nearly everything, including reputation or morality, for other goods, it is easy for misconduct to arise.

2.3 Double Standards

"Some behaviors promoted to further creativity [could be considered] ethically objectionable, such as risk taking, rule breaking, conflict creation, and challenging authority." For some "employees the meaning of 'acceptable behavior' deviates from the generally accepted criteria to the point of becoming fraudulent activity." There may be "various reasons for wrongdoing, such as competitive pressures, ambiguous standards, and intimate relationships between regulators and contractors" (Lopez, 2015, p. 47).

"The 'never mind the rules' disposition of some employees allow them to settle fundamental ethical issues on their own, including which rules to break, under what circumstances, how far to go, and who can do it." Nevertheless, there is an implicit understanding that "some rules should never be broken"

(Baucus et al., 2008, p. 103). These rules are linked to the project manager value judgment about what is suitable behavior to meet its goals.

> *Some rules are more likely to be violated than others when performance falls below aspiration levels. A systematic selectivity for rule violation [in projects] is the result of a performance shortfall shaped by contextual conditions (such as the meaning that employees attach to the project structure), characteristics of the rules (structural secrecy, coupling violations and outcomes, enforceability, procedural emphasis, and connectedness with other rules), and the individual normative cultural perspective. It is also the result of differences between rules and the perception of the risk of violating such rules (content, enforceability, relationships with other rules, and project attention)"* (Lopez, 2015, p. 48).

Rules shape project actions. "Those rules could be internal policies and procedures or laws and regulations. The violation of rules has two sources: (a) individuals or groups acting in their [project] roles and on behalf of project goals, and (b) employee deviance, misbehavior, and misconduct" (Lopez, 2015, pp. 48-49).

Merton (1938) found that social structures exert pressure on certain individuals in society to engage in nonconformist conduct. Two elements of social and cultural structure are important: culturally defined goals, purposes, and interests (frame of aspirational reference), and the acceptable modes of achieving these goals. In every social group there is a coupling between the scales of desired ends and the permissible procedures for attaining them. Findings related to violations of the projects' rules are the direct result of the perceptions and behaviors of individuals in decision-making roles (i.e., project managers).

Manifest and latent functions can be distinguished. "Manifest functions are open, stated, conscious functions; latent functions are unconscious or unintended and may reflect hidden purposes" (Schaefer, 2005, p. 14). Sherron Watkins (Enron's whistleblower) said, "Enron certainly had a strong culture. But our stated values of respect, integrity, communication, and excellence (RICE) did not describe that culture," pointing to the existence of two different sets of values, the stated and the real (which came from the unspoken culture). She emphasized: "At Enron, we had a firm culture in place that emphasized making earnings targets no matter what, and I don't think any one person could have changed that culture" (Beenen & Pinto, 2009, p. 279).

"The existence of this duality of values jeopardized the whole concept of trust. This double standard constitutes a major difficulty in the formulation of [project] codes, frequently regarded by workers as boasting about the way things should be in contrast to the way things are. Employees evaluate [projects] negatively when they perceive a deviation between words and reality, as they value honesty in communications" (Lopez, 2015, p. 50). "If a forbidden pattern flourishes in the project without consequences, that pattern becomes a de facto expectation for project members, sabotaging the restraining influence of official ethics policy" (Ross & Benson, 1995, p. 346). Dougherty (2008, p. 1) wrote: "many employees believe, by all accounts, that bribes were not only acceptable but also implicitly encouraged." Other cases of double standards are the situations of mild abuse consented by some stakeholders that are due to a self-interested rationale to do so.

The general understanding of personal moral values has evolved. Originally, it was believed to be a unique and constant set of values. Later, a concept arose that assigned two sets of values to certain groups (such as businesspeople): one set of ethical values for home and another set for public life. Thus, executive decisions are based on business considerations, and other, non-business-related decisions are based on a private ethical code (Lopez, 2015, p. 51). Ariely (2008, p. 68) said that "we live simultaneously in two different worlds—one where social norms prevail, and the other where market norms make the rules."

As a modern view ponders consciousness and the tolerance of a disorderly world of assorted motives, contemporary arguments posit that people may have various sets of ethical values. "We may, in fact, be an agglomeration of multiple selves" (Ariely, 2008, p. 105).

2.4 Honesty

The moral hazard problem originates in the presumption that, in any situation, humans will act for their own benefit. This implies opportunistic or even unethical behavior. Although individuals might favor honesty, they can be tempted. Even accepting that managers in general act ethically, the likelihood of managerial impropriety increases with the intensity of the motivation. "It is easy to be ethical when a small portion of one's pay is at stake; it is harder to be ethical when substantial portions of one's pay can be influenced through misrepresentation" (Harris & Bromiley, 2007, pp. 352–353).

In the past, honesty was frequently explicated in terms of personal dispositions and character traits. Cases of moral wrongdoing were seen as defects in the person's character. "Resisting temptation and instilling self-control are general human goals, and repeatedly failing to achieve them is a source of much of our misery" (Ariely, 2008, p. 116). A modern approach to honesty gives priority to the circumstances and the environment of decision making, rather than the individual decision maker's moral beliefs and convictions, as it is now understood that human decisions and behavior are significantly affected by aspects of the situation (see Case 2-4).

Case 2-4 Character and Circumstances

*In one experiment designed to explore the balance between charac-
ter and circumstances with regard to helpfulness, theology students at
Princeton University were individually told to walk to another part
of campus to do a presentation on The Good Samaritan story from the
Bible. One-third of the students were told that they needed to hurry up
to get to the building in time, another third that they were just on time,
and the final third that they were early and had plenty of time. On the
way to the other building, the students encountered a person lying on the
pavement in pain, needing assistance.*

*The researches wanted to test whether the differences in the students'
hurry to reach the other building would make a difference in their helping behavior. If character is the most influential factor, then only minor
differences should be observed. In the experiment, only 10 percent of the
students in a hurry offered to help. In total, 45 percent of students who
were on time and 63 percent of those who were early made helping initiatives to the person in pain* (Darley and Batson, 1973, p. 105). *The
results indicate that circumstances have a stronger influence on conduct
than character* (Kvalnes, 2014, p. 594).

Questions

1. How do we assess risk in the face of time pressures?
2. What ethical theory might explain this behavior?
3. Is it wrong for onlookers to "do nothing" while a person in distress needs aid? Explain.
4. Are "Good Samaritan" ethics relevant to all cultures, or are there cultures in which there is no duty to respond?

Dishonesty may be the result of a process where people neutralize initial moral dissonance and find excuses that allow them to act against their own moral convictions. Dishonesty (withholding and misreporting) in projects can be understood as the result of neutralization carried by project members who are normally committed to being honest, but who convince themselves that dishonesty is acceptable under their current situation. In a project context, certain justifications for misreporting are used, tolerated, and even encouraged (see Case 2-5). The hesitation to carry bad news represents a substantial challenge in project management, as it leads to intentional misreporting.

Project members who perceive their projects as those in which rules are strictly enforced are less inclined to misreport, while those who perceive their projects as dominated by self-interest are more inclined to misreport. Project managers tend to be less transparent about project progress in organizations controlled by behavior, where as a result of little flexibility they need to pursue ways around the system. Honesty can be promoted by identifying moral neutralization undertakes and depicting them as unacceptable.

Case 2-5 Timekeeping and Layoffs

In a highly projectized telecommunication company, the timekeeping system was precise to the hour and managed with care by project as well as functional managers.

If the time was expended doing project work, employees entered the project number into a web-based tool. This time was approved weekly by the project manager and its cost booked toward the project expenses. If the time was expended doing other activities such as holiday, vacation, personal time off, medical leave, family leave, jury duty, training, and some other less common activities, employees entered its description into the tool. This time was also approved weekly by the functional manager and its cost booked toward the functional cost center.

In case the employees were not required by any project, they couldn't get approval for a training, and they had exhausted their vacation and personal time off, the last option available was to book that time as "home based." This time was considered overhead.

A "utilization" metric was defined as the ratio between all the times booked in categories other than "home based" and the total available time.

Optimization efforts made yearly looked into this timekeeping system for signals of resource wasting. The system provided an immediate view of which functional areas in general and which employees in particular were at the top of the "home based" category. Perhaps their set of skills was no longer useful in the ever-changing technological arena, or perhaps they didn't have the flexibility to adjust to the available projects, which many times required travel and stays in places other than their home base. In one case or another, this category usually was the first one to be trimmed (despite employee performance), along with the underperformers (despite their utilization).

As a consequence, employees were positively motivated to adopt a more flexible posture, to get up-to-date training in order to keep their competitiveness, and to look actively for projects that could require their expertise. Nevertheless, there was also a backlash of motivation for those facing "home based" situations. They felt like prospective layoff candidates, so they rationalized that any means to survive in the company should be allowed, as their value could be useful in future projects. Some begged friends in project manager (PM) positions to let them book time toward their projects, others begged functional managers to approve unnecessary training, others booked time on random projects expecting that the PMs wouldn't noticed it.

While pressure could positively motivate employees, beyond a certain point it could also jeopardize their ethical posture, when the matters at stake affected their core interests (performance evaluation, promotion, continuity of employment, bonus, etc.).

Questions

1. Could you describe other cases when pressure helped to influence employee performance and when it led employees to compromise ethical standards?
2. In your view, why did the examples you named in Question 1 happen?
3. Would this situation be acceptable if it occurred in a culture where it was ethical?

Müller et al. (2013, p. 35) found that information is usually withheld for the following reasons:

1) *Hoping to be able to balance costs through reduced functionality of the product*
2) *Hoping to recover through other means at some time in the future*
3) *Fearing project termination*
4) *Fearing face-loss by admitting planning mistakes*
5) *Fearing to lose bonuses or other incentives*
6) *Uncertainty about proper timing for escalation*

Kvalnes (2014) found that intentional misreporting is affected by risk propensity and career aspirations, the relationship between the reporter and the receiver of information, and the extent to which there is reciprocal trust between project members and the governance structure.

2.5 Codes of Ethics

In order to guide behavior in the face of ethical dilemmas, bodies such as the Project Management Institute (PMI) and the Association for Project Management (APM) have crafted Codes of Ethics and Professional Conduct. The PMI Code is focused on the necessary processes for compliance, the APM Code is focused on general management aspects, skills, tools, techniques, and context.

> *Codes of ethics are forms of organizational expressions of ethics, laws, and control; they include readings of values, processes, and structures. These documents are contrived and understood within larger social systems of meanings, norms, and authority. Organizations inculcate their core values and their ethical codes through discourse; the code content manifests the thoughts of those outlining the code instead of those belonging to the entire organization* (Lopez, 2015, p. 51) (see Table 2-1).

Table 2-1 Code of Ethics

Think before you act.
Respect all people.
Look after your colleagues.
Strive for excellence.
Compete fairly.
Honor agreements.
Don't break the law.
Don't bend the law.
Be honest.
Stand up for what you believe is right.

> *Codes do not determine behavior since they proceed to conclusions through mental representation of a system that does not promote reflexivity in members of the organization in terms of ethical precepts such as honesty, integrity, responsibility, and respect. In contrast, there is strong consensus regarding the idea that codes are public arguments of organizational values and that they can positively influence the ethical environment* (Lopez, 2015, p. 51).

> *While values, integrity, and ethics are directly foregrounded in the corporate codes of ethics, the majority of the content centers on adopting particular policies, regulations, laws, and practices, considering a social system that places ethics within the sphere of legal demands and compliance to rules, encourages company interests, and respects conventional control of organizational behavior. Such a social system seeks to push ethical behavior through the accentuation of legal requirements, regulatory processes, and formal control of conduct. Codes emphasize laws, regulations, and compliance procedures. The emphasis of codes of ethics on laws and policies deflects fellow [project] members from centering on their own morality, decision processes, and governance structures. These codes illustrate a social system that has a compounded consciousness of ethics in the wake of widespread misbehaviors and governance demands. Thus, codes applied post the Sarbanes-Oxley Act (SOX) highlight the relevance of compliance and law more conspicuously than do pre-SOX codes. Organizations employing auditors are more likely to have a code of conduct and periodical appraisals of internal controls. Although the most influential terms within the codes of ethics involve elements applicable to themes of ethics, laws, and practices, much of the codes represent legal demands and checks of routine work patterns* (Lopez, 2015, p. 52).

The increased interest in institutionalizing ethics comes from a general impression that having an ethics code might protect a project from significant vulnerability to criminal fines (Laufer & Robertson, 1997).

> *In light of these facts, some questions remain:*
>
> a) *To what degree are ethical principles and legal requirements constituted as fundamental concerns in the codes of ethics?*
> b) *Do new legislation and codes of ethics improve governance and business culture, nurturing legal abidance and ethical selections?* (Lopez, 2015, p. 53).

Clarity, comprehensiveness, and enforceability are the qualitative characteristics of a code of ethics. "Only when codes of ethics are consistent with the [project's] culture and when they are enforced, can they effectively impact ethical behavior. Formal conventions become internalized by enforcement and positive predisposition. Formal conventions are thus followed because of the notion of authority and because of positive predispositions induced by education" (Lopez, 2015, p. 53). Nevertheless, it should be considered that "although modest external pressure certainly can stimulate the internalization of virtues, beyond a certain level external pressure may also have a negative effect. It may feed an attitude of minimal compliance and distrust" (Dubbink et al., 2008, p. 393).

Although regulations cannot completely define employee actions, for the reason that they are their individual responsibility, an ethics code is a means of formal social control responding to deviant conduct, which helps to stiffen the natural order in complex projects. An artificial order is created and maintained by business ethics initiatives with the objective of exerting social control. This order is assured by validating norms of satisfactory business practice and behavior. The ethical codes frequently include conditions that involve customs, etiquette, and project norms that are not always required by law (Lopez, 2015, p. 53).

"Codes of ethics are the result of a social system that attempts to encourage ethical behavior by accentuating regulatory procedures, legal requirements, and formal control of behavior" (Canary & Jennings, 2008, p. 277).

"In impersonal modern societies, social order is maintained by adjusting the natural order with artificial constraints." The instinctive bonds, common in times of rural or small-town communities, no longer tie people in the vast and complex present-day metropolis. "Natural and informal social controls (empathy, sociability, and solidarity) became substandard to control and inhibit self-interest. As primitive communities grew into "artificial societies," ordinances to appease enticement demanded a more formal and systematic style. In other words, an artificial order was required" (Lopez, 2015, p. 54). The need for ethics initiatives becomes more important with the size and complexity of the project. As projects become more hierarchical and stratified, there is a demand to codify common principles and expectations (Laufer & Robertson, 1997).

The means of social order control are "public opinion, law, beliefs or values, tradition and convention, education, custom, religion, social roles, ceremony, art, personality, enlightenment, illusion, social valuations, ethics, and class control. According to the view of social control of social bonds, ethics initiatives render a chance for employees to tone their allegiance to their job and comrades. Offending the code of ethics would risk the condemnation of fellow co-workers" Lopez, 2015, p. 54). On the other hand, skeptics have thought that remuneration is the predominant means to control employees, so that neither social bond can be established, nor can social control exist in an individualist society.

Hofstede (1984, p. 92) explained that "the extent to which people feel that behavior should follow fixed rules differs from one culture to another. In cultures high on the uncertainty avoidance scale, behavior tends to be rigidly prescribed either by written rules or by unwritten social codes. The presence of these rules satisfies people's emotional need for order and predictability in society. . . . In cultures low on the uncertainty avoidance scale, there will also be written and unwritten rules but they are considered more a matter of convenience and less sacrosanct. People are able to live comfortably in situations where there are no rules and where they are free to indulge in their own behavior. . . . People here are more pragmatic, even opportunistic."

An analysis of codes of ethics shows disproportional attention given to conduct against the corporation. "The accent of the codes appears to be internal (governing employee behavior) instead of external (managing relations with stakeholders). According to normative control theory, formal controls (such as compliance codes) are more ineffective than natural, informal controls" (Lopez, 2015, p. 55). Artificial order is a sign of an anomic project.

As rules separate acceptable from unacceptable behavior, *they are typically regarded as easier to follow than principles,* as the latter could allow for considerable latitude. Nevertheless, there are arguments that oppose this view; these arguments refer to their complexity, timely update, and the possibility to encircle and avoid them. "The PMI Code of Ethics and Professional Conduct [Project Management Institute, 2012] recognized aspirational and mandatory standards for responsibility, respect, fairness, and honesty. Rules and regulations can be associated with the handling of clear standards to employees. Procedures also provide employees with a sense of community" (Lopez, 2015, p. 56). "Control theory suggests that our connection to members of society leads us to systematically conform to society's norms . . . our bonds to family members, friends, and peers induce us to follow the mores and folkways of our society. . . . Socialization develops our self-control so well that we don't need further pressure to obey social norms (Schaefer, 2005, p. 176).

A project that dictates its code of ethics coming from top management will struggle to promote the participation of all employees. If the target is to manifest due diligence in the case of a criminal indictment, project managers should not expect a strong employee commitment. An overreliance on formal control may reflect a project with an autocratic and coercive culture.

As observed by Laufer & Robertson (1997, pp. 1029–1031), in general it can be said that:

- *The implementation of ethics codes and programs has taken place without a clear theoretical foundation.*
- *The effectiveness and impact of ethics codes are rarely examined.*

An increased interest in the institutionalization of ethics came as result of the general impression that an ethics code or program could help decrease exposure to criminal fines.

It has also been suggested that codes of ethics should be implemented to enhance companies' reputation. In this regard, the U.S. Department of Justice, in a memorandum sent to U.S. Attorneys with respect to the principles of federal prosecution of business organizations, highlighted that: "Prosecutors should therefore attempt to determine whether a corporation's compliance program is merely a 'paper program' or whether it was designed and implemented in an effective manner" (Thomson, 2003).

The project code of conduct should:

1. Establish the sponsor commitment to the code of conduct, stating the objectives of the project, its values, and the expectations of all stakeholders.
2. Identify measures the project adopts to encourage the reporting of unlawful or unethical behavior and to actively promote ethical behavior, and describe the means by which the project monitors and ensures compliance with its code.
3. Detail the project's responsibilities to stakeholders. This might include reference to the project's reporting policies, practices, and disclosure, as well as to product quality.
4. Describe the project's approach to the community, including support for community activities and environmental protection policies.
5. Include the project's privacy policy and how the project handles conflicts of interest.
6. Depict the project's employment practices, occupational health and safety, training, and substance abuse policies.
7. Describe the project's approach to bribes, courtesies, facilitation, incentives, and commissions, stating the standards followed to encourage compliance with legislation.

2.6 Summary

Ethics was not examined here from a top-down or aprioristic perspective where values and conducts were explicitly prescribed; instead it was considered as the observed choices made in practice by ordinary agents assigning value according to social common sense, in a bottom-up, value-free descriptive approach (Lopez, 2015, p. 71).

There are two reasons for [projects] to engage in ethical practices. First is the ethical motivation to do the right thing; in this case no outside pressure or governmental constraint is required, as project members choose this approach according to their own will, recognizing themselves as members of society. Second is the Machiavellian approach; in this case conviction is lacking, outside pressure and governmental constraints are the motivations for actions that aim to show that a project is doing the right thing. The ulterior purpose of these actions is to avoid legal consequences and/or to deceive the stakeholders and general public by making them believe that the project prioritizes ethical practices (Lopez, 2015, p. 72).

People believe in an orderly world where everyone gets what they deserve, where good is rewarded and bad is punished. In this world, loyalty, commitment, job satisfaction, and trust in management are influenced by the perception of fairness, and projects are able to prevent ethical lapses by a periodic reshaping of its standards to maintain the ethical goals aligned with those of the society.

There is a need for awareness regarding ethical issues that applies to projects, and this awareness ought to be embedded into project management practices. Ethics, like quality, cannot be added at the end of the project—it must be part of every single project process.

Chapter 3

Context

Context refers to the spatial, temporal, and situational dimensions that influence the state of affairs. It is the combination of all circumstances at a given time and place. It encompasses a wide range of variables, including national values and project culture, as well as the global, regional, national, and local realities of the time. As such, context is essential to define what is to be considered and interpreted, as it has biasing effects for some senses. Sherron Watkins said that "the context you're in can be a powerful influence on whether you view something to be right or wrong" (Beenen & Pinto, 2009, p. 280).

In all social constructions, context influences the decision of what values are accepted. It can be observed that different societies or groups have dissimilar values according to their contexts. The importance and influence of context for projects is manifested in the cultural, political, and business environment in which the projects operate. Context is recognized in *The Handbook of Project-Based Management* (Turner, 2009) as one level of governance within project-based organizations.

Along with the economic detachment from moral concerns, a shift in the meaning of morality also has occurred. With a context that could include a loosely regulated environment, lax oversight, huge rewards, and trivial penalties, it is not uncommon to expect project misbehavior.

Context is also embedded in the definition of the stakeholders, as they are the individuals or groups who can significantly impact, or be impacted by, the project (including employees, customers, communities, and government officials). Project managers' decisions that affect the stakeholders also modify the project's context.

As the development of meanings derives from social interactions, it is affected by the context. Sociologists have used the term *Hawthorne effect* (also *observer effect*) when referring to subjects of research who deviate from their typical behavior because they realize that they are under observation. According to the observer effect, the behaviors of employees become unnatural and biased by the context in which they are immersed or in which they think they are immersed.

3.1 Perception

Perception and other related concepts such as assumption, belief, impression, understanding, construct, judgment, assessment, indication, regard, view, subjective morality, and subjective mental models, are also associated with context. As people are conditioned by education and life experiences, it is common to find differences in perceptions of the same reality. Public opinion is often based on impressions and perceptions rather than conscious study.

A *perception* is a mental representation formed upon the identification and interpretation of a sensation; it can be influenced by learning, memory, motivation, and expectations. Our perceptions of the social environment are made of diffuse information, various beliefs, dissimilar motivators, and expectations, affected by subjective biases and predisposition. Experiences help create mental representations of what is known and what to expect from a particular subject, biasing or predisposing perception in a certain way.

As legality does not invariably involve the totality of an issue's perceived morality, there can be separated the objective and the subjective moralities. Objective morality is the generally agreed-on societal moral law, which corresponds with promulgated laws. In contrast, subjective morality is the individual's belief regarding the appropriateness or incorrectness of an action, which corresponds with the concepts of moral sense, scruples, and conscience. Subjective mental models, such as the project culture, are relevant because people employ pattern recognition when facing uncertainty. Then, project culture is the expectations of conduct that act on the members of the project.

Legitimacy is the general perception that activities are proper and desirable inside a socially fabricated system of norms, values, and beliefs. The core beliefs and principles that are viewed as desirable are recognized as values; they are derived from the individual's membership in a culture. Attitudes, beliefs, and behaviors make a continuous spiral of community culture. People rely on values as a way to rationalize their behavior. Values specify an individual's personal beliefs about how to behave; intrinsically they are modes of behavior. As products of a culture, their attachment to social facets is manifested whenever individuals feel guilt when acting inconsistently with endorsed social expectations.

The substance of values varies from culture to culture. Society determines what is right and what is good, and as different societies choose differently in what they pronounce as right or wrong, it is difficult to equate the right with the social and the legal. However, there is a perception that beneath the decrees of any given society rest moral principles and ecumenical moral rights, and the cogency of any moral selection leans on the principles that selection substantiates. Such moral principles are universal and make up a feasible standard against which the laws or rules of any society should be evaluated. For instance, justice is a basic and universal standard.

Instead of coming from rationality and logic, the standards of conduct follow cultural conventions. Consequently, managing a project is culturally dependent.

Some believe that reality is not universal but plural, asserting that there are multiple realities, and each one is based on different assumptions. Neither harmony and consensus nor power and conflict can provide the complete picture. The drivers of sociological perception are dynamically related in time and space, which makes the finding of sense for the social order problematic, with no simple recipe available for every situation. In a multivariable environment, having variables in excess of processing capacity could be defined as chaos; therefore, sociological perception is in the borderline between order and chaos, depending on the complexity of the situation and the capacity for understanding its variables. It can be inferred that perception of values is affected by four superimposed drivers.

1. The *global driver* accounts for the global sense of values. It changes slowly over time (centuries or decades) and applies to wide geographic areas such as continents or blocks of nations—for example, the Medieval driver (which lasted from the 5th to the 14th centuries, characterized by obscurantism and feudalism) and the Renaissance driver (which lasted from the 14th to the 17th centuries, characterized by humanism and a flourishing of the arts).

2. The *local driver* accounts for the national or local sense of values. It changes over years (possibly from one government to another) and applies to geographic areas ranging from a town to a number of countries—for example, the New Deal driver (domestic programs enacted in the United States between 1933 and 1938, aiming for relief, recovery, and reform) and the Eurozone driver (political-economic union of European states).

3. The *industry driver* accounts for a specific subjective sense of values. It changes as economy, technology, and other factors evolve—for example, the steam power driver (the motor of the industrial revolution from the earliest 1700s), the electricity driver (the great progress in electrical

engineering since the late 19th century through the early 20th century), and the IT driver (since the advent of the transistor and the computer).
4. The *project driver* accounts for a project's sense of values. This changes with internal policies and processes.

Making a parallel with the Heisenberg[*] uncertainty principle, which established that uncertainty is actually a property of the physical world, it can be said that when matters depend on sociological perceptions, uncertainty plays a key role in the final outcome.

Values, integrity, ethics, and responsibility are demanded in the contemporary project workplace, because of a growing acknowledgment that good ethics is a socially shared value. The perceived quality of a project's ethics can influence the cost of raising capital, making a case again about the importance of perceptions. As such, the project manager's and the project team members' engagement with society must be conducted ethically, and managed according to values, as isolated measures of profitability fail to catch the substance of a project's performance.

The objectives of project ethics initiatives are multiple and directed to exert influence on the external and internal constituents' perception of the project, the project culture, and the employees' behaviors and practices. The perceptions of project ethics are determinants in establishing the foundational elements of a code of ethics. Then, project managers will likely incorporate this code of ethics into their decisions when pressure to do so is perceived from stakeholders instead of regulatory agencies, and when they believe that a positive project culture is promoted.

Although there is a widespread perception that irresponsible project behavior exists, general impressions of corruption and fraud are frequently biased by individuals' sense of discontent with their own lives. Trust in government and business decreases when unemployment grows. Thus, after years of gloomy employment figures, distrust is disseminated widely, affecting a range of institutions including the government, the judiciary system, labor unions, businesses, schools, and religious institutions. Most people seem to believe that corruption has become more prevalent in recent years, although others believe that corruption may have been a constant through the years, it is just easier to be aware of today because of our broad access to news and social media.

A variety of contextual variables influences the perception of the project culture, and how ethical dilemmas are addressed.

---

[*] Werner Karl Heisenberg (1901–1976), German theoretical physicist. The uncertainty principle (1927) is also known as the indeterminacy principle.

3.2 Motivation

Motivation refers to the direction, intensity, and persistence of voluntary, goal-oriented behavior. We define the motivation primal statement as a person's wish to have an adequate job, a decent place to live, and to be engaged with the community.

Many projects managers implicitly understand the motivation primal statement. They realize that it involves concerns about safety, ergonomics, compensation, work environment, health services, and pension plans. Some go beyond by building neighborhoods for their employees, fostering clubs, civil associations, and other community activities. However, after Friedman (1970, p. 124) postulated that "there is one and only one social responsibility of business—to use its resources and engage in activities designed to increase its profits so long as it stays within the rules of the game," the adherence to the motivation primal statement lost representation in the mainstream, being confined to a few projects heavily linked to their communities and to some not-for-profit areas (such as people working in the military). For most other projects the attention shifted to job motivation, which is really only one of the motivational drivers described by the motivation primal statement.

Three main factors affect employee job motivation: the organization, the job itself, and the leader. These factors are limited to the *adequate job* driver, and they make no reference to *a decent place to live* or *to be engaged with the community.*

The dynamics of individual needs are explained by the content theories of motivation (see Figure 3-1), which recognize those needs as deficiencies that ignite behaviors toward their fulfillment. As such, unsatisfied needs are the roots of motivation.

The first content theory was Maslow's (1943) "Needs Hierarchy Theory." Maslow set out a hierarchy of five levels of human needs. From the bottom, this hierarchy starts with physiological needs (food, air, and water); at the second level are safety needs (shelter and a safe environment); the third level is occupied by the need to belong (love, affection, and interaction with others); the fourth level is for the need for esteem (personal achievement, social recognition, and respect from others); finally, at the top of the hierarchy there is the self-actualization need (self-potential realized).

Although an individual's behavior is motivated by several levels of needs simultaneously, Maslow stated that the lowest unsatisfied need at the time is the primary behavior motivator. Once this need is fulfilled, the next need in the hierarchy takes the place of the primary behavior motivator, in a process known as satisfaction-progression. In this way, an individual behavior's first primary motivator always comes from the bottom level (physiological). Only after the

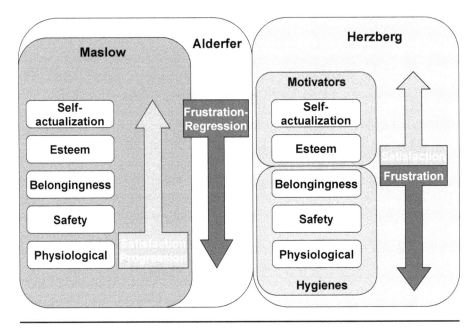

Figure 3-1   Content Theories of Motivation

needs at this level are fulfilled does a primary behavior motivator move to the next level (safety), and so on.

Researchers have found this model too rigid, as needs are unstable and dynamic, and there is rarely a complete match for these levels. A later approach by Alderfer (1972) added a process known as frustration-regression, which states that when a higher level cannot be satisfied, the primary behavior motivator moves back to a lower level.

Another contribution to this field was made by Herzberg (1966), who, instead of suggesting that people change their needs over time, proposed that growth and esteem needs are everlasting motivators related to satisfaction, while in contrast, there are needs called hygiene factors (physiological, safety, belongingness), whose lack is related to dissatisfaction (frustration). Under this approach, there is a coexistence of satisfaction and frustration feelings, each affected by a different set of needs. An improvement in the lower levels reduces frustration but does not generate satisfaction. Likewise, an improvement in the upper levels generates satisfaction but does not reduce frustration.

It is not difficult to find a fit between the motivation primal statement and the content theories of motivation. However, when looking at employee motivation, it seems to cover a narrow scope almost limited to the hygiene factor needs. The first consequence of Friedman's (1970) postulate was a decrease in the scope of employee motivation to a 40-hour, in-house labor week (see

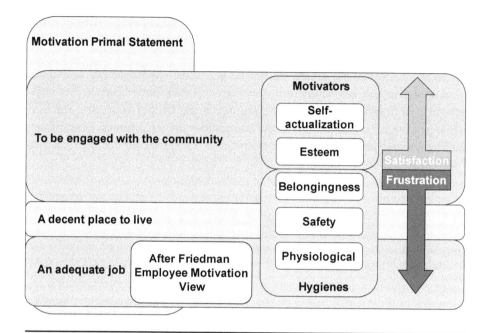

Figure 3-2   First Consequence of Friedman's Postulate on Motivation

Figure 3-2). This motivation can only mitigate frustration, but a wider scope including satisfaction is not possible when one is acting exclusively at the level of the needs of hygiene factors.

A second consequence of Friedman's (1970) postulate subtly affected the definition of motivation itself (see Figure 3-3). Motivation was defined as the

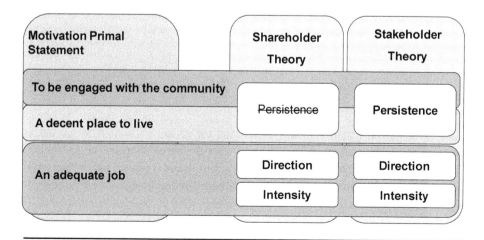

Figure 3-3   Second Consequence of Friedman's Postulate on Motivation

direction, intensity, and persistence of voluntary, goal-oriented behavior, but persistence is a quality related to loyalty and pride, which are nurtured when individuals feel safe and belonging to the community. This is a long-term characteristic of motivation, which cannot be generated spontaneously as a result of extrinsic or intrinsic short-term incentives. A lack of persistence is a characteristic of the shareholder theory common in the United States and the UK, while persistence is still present in the stakeholder theory common in continental Europe and Japan. This characteristic affects the measurements of success, which should necessarily be short-term for the shareholder theory, limiting its scope to yearly financials, regardless of risks taken, and with no consideration of other long-term indicators such as generation of goodwill and business survival.

In consequence, the reduction in the scope of motivation resulting from the general adoption of Friedman's (1970) statement resulted in employee frustration, which facilitated the violation of norms.

Opposing Friedman's statement, new ideas were later introduced, trying to create value for all the project's constituents (investors, customers, employees, suppliers, and the community) by considering factors such as customer satisfaction, team excellence and happiness, improvement in the state of the environment, and community support.

A set of motivators and hygiene factors is usually involved in a project environment, resulting in a compound satisfaction/frustration output. As motivation is one of the conducting factors in achieving project objectives, the ability to motivate the team is one of the key attributes of successful project sponsorship.

The lack of motivation or the presence of frustration in the project workplace may drive the violation of norms (workplace deviance), which differs from ethics (right or wrong behavior in terms of justice, law, and morality). "Even good people are not immune to being partially blinded by their own minds. This blindness allows them to take actions that bypass their own moral standards . . . motivation can play tricks on us whether or not we are good, moral people" (Ariely, 2008, p. 227).

3.3 Culture

The study of the responsibilities demanded for social affiliations is known as *deontology*. A deontological view emphasizes the need to preserve individual dignity (inherent worth) in work practices. Emphasis is placed on the personal and ethical context of behavior and the collective moral obligations of subsidiarity, solidarity, and justice.

Culture is frequently assumed to include shared basic assumptions, customs, myths, and ceremonies that convey fundamental beliefs, and it is manifested by values expressed in individual and group behaviors. Culture constitutes the parameters of acceptable and unacceptable behaviors, establishing recognized and accepted premises for decision making. As such, culture should not be associated exclusively to different countries or regions. There are other types of culture, for example, company, industry, professional, functional, and even project-level cultures.

Our first notions of ethics come from parental example. Socialization introduces additional ethical concepts, religion provides moral teachings, and society at large shows what are acceptable and unacceptable behaviors.

Beliefs and worldviews are molded by the culture in which people grow up, the legal system provides the substrate to ensure order and stability, while hierarchy recognition as well as respect for authority develop later. From this, it can be concluded that ethics is influenced by religion, culture, law, politics, tradition, and many other social factors. It is made up of both universals and local beliefs.

Two elements of the social and cultural structure are worth considering here: (1) culturally defined goals, purposes, and interests that comprise a frame of aspirational reference; and (2) the definitions, regulations, and controls imposed by the social structure when describing the acceptable modes of achieving these goals. Social groups match their scale of desired ends with the moral or institutional regulations that provide permissible ways of achieving these ends. Even though cultural goals function together with institutional norms, the areas of possible behaviors and purposes do not bear a one-to-one relationship to one another. In most cases, social pressure cannot be overcame by high morals or professionalism.

Anomie is defined as the lack of moral standards in a society, and also as personal state of isolation and anxiety resulting from a lack of social control and regulation. It is ethically desirable in a social structure to have a coordinated association between goals and the recognized means of attaining those goals. Lack of this association results in anomie, which is evident when some individuals engage in nonconformist rather than conformist conduct as a result of pressure exerted by social structures. This kind of ethical misconduct corrupts the spirit and integrity of the project culture. As today's lifestyle foments rampant individualism, encourages frantic consumerism, reduces moral prejudices, and adopts an ad-hoc morality, there is a critical need to restore ethical conduct.

Project culture can be defined as the expectations of conduct and the collective mindset that influence the members of a project. In this sense, expectations of conduct and collective mindset are closely related to the context in which the project is being conducted, and with the perceptions and the motivations of the project manager and the project team members.

Some cultural variables are recognized as potential sources for uprising against external ethical standards:

- A risk taker's reward structure that is careless of ethical limits
- A rationalization that "everyone plays the game"
- Ambiguous ethical rules paired with deficient training
- Inability to comprehend that ethics is a legitimate constraint of projects
- No concern about sanctions for ethical violations

The stiffer the achievement orientation, individualism, and materialism, the heavier will be the willingness to justify ethically suspicious behaviors. In such a situation, cultural values stress achieving materialistic and economic ends over the legitimacy of the means applied to attain these ends.

As anomie describes the loss of direction (state of normlessness) when social control of individual conduct became ineffective; *deviance* is the conduct that violates the standard of behavior. It involves the infringement of group norms, including not only criminal conduct but also many activities that are not matters for prosecution. Deviance is not objective; rather, it is a matter of social definition within a particular society and at a specific time. It should not be referenced to an absolute system of moral standards, but to the standards of a specific social group.

Employee deviance is voluntary conduct that violates significant norms, thereby jeopardizing the welfare of the project and/or its members. By referring to significant norms, this definition specifically excludes minor infractions of social norms that are not harmful. Examples of significant infractions are theft, bribery, quality violations, harassment, false reporting, and safety violations, among others.

In some cultures, punishment helps to fix acceptable behavior, thus contributing to stability. When wrong behaviors are not sanctioned, people may modify their standards of appropriate conduct. The deterrence theory, and the social learning theory, provide explanations of the reasons why individuals refrain from deviance and embrace conformism. Deterrence theory establishes that the perceived risk of punishment deters misconduct, suggesting that characteristics of the sanction, such as certainty and severity, influence deterrence effectiveness. The deterrence of misconduct is associated with a certain punishment severity threshold. It is also observed that the expectation of informal sanctioning from the group to which the individual belongs acts as a powerful deterrent to misconduct.

Six value dimensions, according to Hofstede (1984), represent elements of common structure for any cultural system:

1. *Individualism/collectivism* refers to the orientation of the decision making, the achievements, and the way of working—whether they are revolving around the individual or around the group.

2. *Power distance* is about the extent to which the less powerful members accept and expect an unequal distribution of power.
3. *Uncertainty avoidance* refers to a society's leeway for uncertainness and equivocalness. It also gives an indication of the change propensity or the strife for stability.
4. *Masculinity/femininity* focuses on the degree to which a society emphasizes accomplishment or fostering. Masculinity accentuates ambition and differentiation. Femininity stresses caring behaviors, equality, and awareness.
5. *Long/short-term orientation* refers to the focus of people's efforts, which could be in the future (long) or in the present (short).
6. *Indulgence/restraint* is related to the gratification or control of desires.

Cultural universalism or convergence was suggested after finding that project managers from different national cultures describe similar project practices as ethically suspicious. In this case, those project managers share hyper-norms accepted by all cultures and projects, although the degree of enforcement of such hyper-norms and justification for their violation may vary around the globe or from one company to another. The cultural perspective also applies to the own culture that the organizations develop or to the cultures related to professions. Some people support the idea that ethical principles are objective commands that overstep countries, religions, and times, rather than being subjective matters that vary with cultural, social, and economic conditions. In contrast, others have found national differences in rates of deviant and illegitimate outcomes that may be due to differences in social institutions and national culture. Although cross-national ethical issues focus mostly on national cultures, other factors should also be considered, such as social institutions and economic ideology.

Strategic challenges arise in multinational projects because of the diversity of moral philosophies across international borders, including varying legal environments and cultural restraints. Even inside a nation, when considering different states, subcultures, and socioeconomic levels, generalizations should be avoided. Something that is ethically correct for work in one organization might be against the code of conduct in another organization. Ethical standards and moral intuition are neither standard nor intuitive.

Cultural relativism and convergence are described as providing insight into global variations in morality. From cultural relativism, it is inferred that ethical standards vary across cultures, and on the opposite side, universalism supports the idea that there should be a unique, global, moral standard and shared values. Convergence suggests changing of standards and norms toward common globally accepted values, while the divergence perspective maintains that culturally unique values will be preserved despite the power of external influences.

In order to understand the disposition of project managers toward behaviors generally considered ethically suspicious from a cross-national perspective, the institutional anomie theory* can be used, considering variables such as achievement, individualism, universalism, pecuniary materialism, economy, polity (degree of welfare socialism), family, and education.

Looking at Hofstede's culture dimensions and analyzing how they apply to projects, we can see that project governance practices vary substantially from one place to another. Power distance and uncertainty avoidance dimensions are highlighted as the only workable predictors of the degree of project governance. In low-power-distance cultures, where people can disagree with authorities and where power is distributed among members, stakeholders may pressure project managers toward practices that improve their governance. A typical example is the contrast between democracies (low-power-distance societies), where people (stakeholders) can influence practices that improve governance, and authoritarian regimes (high-power-distance societies), where people's opinions are ignored, at best.

Uncertainty avoidance is the level to which the members of a society feel uneasy with uncertainty and ambiguity; it is related to formalization of social structures with which people feel comfortable. Rigid codes of beliefs and behavior are maintained in strong uncertainty-avoidance cultures, which are also intolerant of deviant ideas; in contrast, a more relaxed atmosphere is typical of weak uncertainty-avoidance cultures, where tolerance is common and practices prevail over principles. Once again, a comparison between democracy and authoritarianism provides an example that can be extended to the handling of changes in projects.

The nature of project management skills is culturally specific; a technique or doctrine that is appropriate in one culture may not be appropriate in another. The dominant values of a culture are the building blocks of projects.

Collective patterns of thinking, the meaning attached to life, world, roles, values, beliefs, art, and the differentiation between good and evil, true and

---

* Institutional anomie theory, rooted in Durkheim's (1893/1964; 1897/1966) sociological theory of anomie, is used to explain rates of crime and deviance grounded on specific social institutions and cultural values. Durkheim observed that changes in culture lower social controls, resulting in the weakening of norms (anomie) and increased deviance. Later, Merton (1949/1968) explained that when a social system prevents goal achievement, deviant means are likely to be selected to achieve an end. Later, Messner and Rosenfeld (2001) suggested that Merton did not place enough emphasis on the institutional drivers of anomie. As enabling rational detachment from traditional rules and norms increases the disposition for no moral scruples about the means to be chosen to achieve the goals, institutional anomie theory identifies factors that spread egoistic rather than principled ethical thinking in society by contemplating the social institutions and the cultural values.

false, beautiful and ugly, are components of a culture, so project management activity is not culture-free. In different cultural environments, project management activities ought to be adapted or will risk almost certain failure. A clear example occurred a few years ago when a Japanese tire manufacturer, in an effort to diversify its production of giant tires, which had been concentrated in only one factory in Japan, decided to install a second factory in Aiken, South Carolina. Planning for this project was made in Japanese style, while execution ran American style. As the leadership failed to identify the cultural differences, the gap grew with time to become a bubble that finally popped, leaving over $200 million in cost overruns, a six-month schedule delay, and a serious quality issue that further delayed the start of operations at the new plant.

The notion of shared goals and collective action has been challenged as unrealistic in the context of large and multinational projects (see Case 3-1).

Case 3-1 Culture and Context

A Swedish engineer once told us:

I worked 10 years ago for an American supplier of mechanical components. We were involved in a project with Nokia for their first 3G mobile phone. One night we were out for dinner in China with a potential Korean supplier for my project. We had already chosen another supplier for this component, but this company was very keen to have Nokia as the end customer. We were three people who had dinner together—me and the purchasing manager from my company, and a representative from the Korean company. This was just before Christmas. The dinner itself was completely normal. We broke up outside the restaurant and I had a short distance to walk to the hotel. When we had shaken hands, the Korean representative took out two packages from his coat and said to us: "By the way, Merry Christmas." I was totally unprepared but received the package and started to walk.

When I got back to the hotel, I opened the package. It contained a brand new iPod, which at that time cost about 2500 SEK. I realized this was beyond the limit considered as a permissible Christmas gift and wondered how I should proceed. The trip back to Europe was scheduled for a few days later, but I was not supposed to meet this supplier again. Should I contact him afterward and point out the impropriety? Or just keep the gift?

I decided to keep the gift. My son got it later as a Christmas present. The supplier never got the job. What the purchasing manager did with the gift I still do not know today.

Questions

1. In cases like this, how should you proceed?
2. Should you have asked the supplier to wait in the street until you opened the present and checked its value? Explain.
3. Should you have contacted the supplier afterward? Explain.
4. Should you keep the present and forget about the incident? After all, if the supplier did not get the contract, your decision was not affected by the gift.
5. Should you notify your supervisor or ethical officer and ask for guidance?
6. In general, how should one handle gift giving when doing business abroad?

3.4 Summary

As moral actions can be influenced heavily by situational variables, the nature of project governance can be captured by saying that project governance practices evolve in the light of the circumstances, complexity, and changes, adapting to the specific context.

The scarcity-munificence of a project's environment influences the commission of illegal acts. The more scarce the project's environment, the more effort the project manager will exert to obtain resources from that environment, and in doing so, the more likely she or he is to utilize means considered unfair market practices, including carrying out legally questionable activities and the plain commission of illegal acts. In contrast, when the environment is munificent (very generous), the project manager is likely to find ample resources without resorting to illegal activities.

The constantly changing expectations of society are a challenge for a project's ethical principles and project management competencies because it is difficult for project managers to know how they should react in the face of the challenges. New developments now frame the context within which projects are reporting, among them a wider demand for transparency and accountability, expanded expectations of project governance, renewed commitment to ethics, calls for a holistic picture of the health and stability of a project, including risk management practices, value creation, and substantial discussions about regulation.

Chapter 4

Governance

Even before ancient Greece and the Roman Empire, different cultures have developed governance structures according to their social circumstances. As time passed, insider or outsider realities moved the pillars of those cultures toward new forms of organization, such as the tyranny or the republic. Many different forces, such as economy, technology, social changes, religion, and military threats, among others, led the change of these governance structures.

The term *governance* developed from the Greek verb *kubernân* (to pilot or steer), used by Plato in reference to the design of a system of rule. It evolved into the Medieval Latin *gubernare,* with the same connotation of piloting, rule making, or steering. It determines who is permitted to make decisions, what are the rules and principles for making the decisions, and how we ascertain compliance with the process. Typical means of governance are regulations, economic means, and information. According to the Global Development Research Center (2015), "good governance has 8 major characteristics. It is participatory, consensus oriented, accountable, transparent, responsive, effective and efficient, equitable and inclusive, and follows the rule of law."

Contemporary governance has its roots in Foucault's neo-liberal philosophy, in which subtle forces in society replace the direct steering of individuals by their supervisors. The ideal calls for self-regulated relationships among the forces within a society, which has shifted toward a less self and more regulated contemporary governance model. As such, governance is a complex phenomenon with many dimensions that must be placed in context and adapted to the circumstances. It involves three main characteristics: legality,[*] legitimacy,[†] and participation.[‡] It may include the following components:

- *Authority structure:* Attaches decision-making responsibility to positions.
- *Mechanisms:* Processes and methodologies required to implement decisions.
- *Policy:* Dissemination of information.

The World Bank (2012, p. 1) states that "governance consists of the traditions and institutions by which authority in a country is exercised. This includes the process by which governments are selected, monitored and replaced; the capacity of the government to effectively formulate and implement sound policies; and the respect of citizens and the state for the institutions that govern economic and social interactions among them."

The World Bank (2012) also considers six dimensions of governance: voice and accountability, political stability and absence of violence, government effectiveness, regulatory quality, rule of law, and control of corruption.

As result of the widespread realization that successful development is related to effective governance, consistent efforts have been conducted in order to assess and measure governance through the use of internationally comparable indexes such as the World Governance Index (WGI), proposed by the Forum for a New World Governance (2012). Indicators are used to provide information, to issue warnings, and to enable action and guidance. They are grouped into five fields: peace and security, rule of law, human rights and participation, sustainable development, and human development.

A more comprehensive review of the contemporary governance includes the following items:

- *Meaning:* piloting, rule making, steering capability, exercise of authority, capacity to formulate and implement policies, management, ordered rule, and collective action
- *Means:* regulations, economic means (economy and employment, basic socioeconomic parameters), access to information, enforcement, traditions, institutions, processes, procedures, methodologies, policy, efficient allocation of resources, structures of authority, coordination, and control, security, integration, expectations, international cooperation, institutional learning, political stability, absence of violence, government effectiveness,

---

[*] Quality of conforming to law. Lawfulness.
[†] The exercise of power is linked to a mandate from the parties involved.
[‡] Act or instance of participating by having a role and sharing responsibilities.

hierarchical coordination, regulatory quality, civil rights, human rights, and control of corruption
- *Characteristics:* moral purpose, legality (rule of law), legitimacy, participation, voice, accountability, sustainable development, transparency, responsiveness, consensus oriented, equity, and inclusiveness.

4.1 Corporate Governance

Corporate governance encompasses relationships among stakeholders, objective definition, strategy selection, corporate structure, communications, shareholder and stakeholder rights, disclosure of information, setting expectations, resource allocation, performance monitoring and control, and the moral purpose of a corporation. It can be thought of as a broad entity that covers the management processes and policies, laws, traditions and institutions affecting authority, and accountability. As such, corporate governance is a complex and holistic subject related to economic efficiency and stakeholder welfare.

A principle of corporate governance proposed by the Australian Securities Exchange Corporate Governance Council (2007, pp. 21–22) is to promote ethical and responsible decision making. In this regard, it said that "companies should not only comply with their legal obligations, but should also consider the reasonable expectations of their stakeholders including: shareholders, employees, customers, suppliers, creditors, consumers and the broader community in which they operate."

As every stakeholder adds some form of value and expects some form of retribution, aligning their different and sometimes competing interests into forms of productive collaboration is not easy, making corporate governance a dynamic force that evolves with changing circumstances.

Corporate governance is associated with strategy, organizational structure, roles and responsibilities, decision making, risk management, policies, accountability, performance, control, and reporting (transparency). Other perspectives also include trust, social capital, and sustainability. Within corporate governance there is the responsibility of business to society: business ethics, corporate social responsibility (CSR), ideology/attitudes/values, corporate social performance, stakeholder management, social issues (treatment of minorities and women, consumers, employees, environment), and business–government relations. It can be thought of as containing the following elements:

1. An arrangement by which organizations are controlled and directed
2. The relationships among senior management, boards of directors, shareholders, and other stakeholders

3. The set of objectives, the means of attaining them, and a way to monitor performance
4. The safeguards (structural, procedural, and cultural), designed to ensure long-term survival

The definitions of corporate governance tend to fall into two categories. The first centers on the conduct of corporations, including performance, efficiency, growth, and treatment of stakeholders (principles); the second relates to the normative framework, including rules under which firms operate—the legal system, the judicial system, financial markets, and labor markets (processes and behaviors). In this way, corporate governance is viewed as a blend of law, regulation, and voluntary practices, which enable the corporation to perform efficiently while respecting the interests of stakeholders and society as a whole (see Table 4-1).

There are two prevalent corporate governance theories: the shareholder theory and the stakeholder theory (see Table 4-2). The shareholder theory, common in Anglo-American countries, prioritizes the interests of shareholders, focusing narrowly on the short-term return on investment (ROI), diminishing or even disregarding other, qualitative objectives. While shareholders are the firm's owners, managers act as their agents. The stakeholder theory, common in continental Europe and Japan, stresses the importance of other stakeholders such as the workers, managers, suppliers, customers, and the community at large, while pursuing long-term objectives related to the creation of value for these stakeholders. According to this view, there is an ethical responsibility to balance the interests of various stakeholders such as consumers, employees, lenders, investors, and government, among others.

Each of these corporate governance theories has its own inherent strengths and weaknesses.

The market-based shareholder theory, characterized by dispersed ownership and the primacy of shareholder value, is the dominant force in international corporate governance. The principal/agent problems are assumed to be paramount. This model has contributed to the dynamism and growth of the United States and other economies that adopted it. However, it has proved not to be exempt from failure, as a number of large corporations have demonstrated, forcing the U.S. government to enact the Sarbanes-Oxley Act (SOX) to drive transparency and disclosure.

In Europe, a relationship-based stakeholder theory has prevailed, indicating a different corporate history and values. It is more dependent on loans from banks than from capital raised in the equity market, and tends to have the support of tight business networks.

Although both theories remain in practice, reflecting fundamental differences in how the values and objectives of corporations are interpreted, the notion of shareholder value is progressively spreading across all parts of the economy.

Table 4-1 Corporate Governance

Source	Definition
Securities and Exchange Board of India Committee for Corporate Governance (SEBI, 2003)	"[Corporate governance] is about commitment to values, about ethical business conduct and about making a distinction between personal and corporate funds in the management of a company."
Sweeney (2008, p. 118)	"Corporate governance are the variable and complex forces, which include organizational structure, legal system, accounting protections, and political structure of the country."
United States Mission to the Organization for Economic Co-operation and Development (2015)	"Corporate governance is the system by which business corporations are directed and controlled. The corporate governance structure specifies the distribution of rights and responsibilities among different participants in the corporation, such as the board, managers, shareholders and other stakeholders, and spells out the rules and procedures for making decisions on corporate affairs. By doing this, it also provides the structure through which the company objectives are set and the means of attaining those objectives and monitoring performance."
Organization for Economic Co-operation and Development (OECD, 2004, p. 11)	"Corporate governance involves a set of relationships between a company's management, its board, its shareholders, and other stakeholders. Corporate governance also provides the structure through which the objectives of the company are set, and the means of attaining those objectives and monitoring performance are determined."
Australian Securities Exchange (ASX, 2007, p. 2)	"Corporate governance is the framework of rules, relationships, systems and processes within and by which authority is exercised and controlled in corporations. . . . It encompasses the mechanisms by which companies, and those in control, are held to account. Corporate governance influences how the objectives of the company are set and achieved, how risk is monitored and assessed, and how performance is optimized. Effective corporate governance structures encourage companies to create value, through entrepreneurialism, innovation, development and exploration, and provide accountability and control systems commensurate with the risks involved."
World-Bank (2012, p. 3)	"In its broadest sense, corporate governance is concerned with holding the balance between economic and social goals and between individual and communal goals. The governance framework is there to encourage the efficient use of resources and equally to require accountability for the stewardship of those resources. The aim is to align as nearly as possible the interest of individuals, of corporations and of society."
Tricker (2000, p. xxi)	"Corporate governance. . . does not have an accepted theoretical base or commonly accepted paradigm. In the words of Pettigrew (1992), corporate governance lacks any form of coherence, either empirically, methodologically or theoretically, with any piecemeal attempts to try and understand and explain how the modern corporation is run. . . . the metamorphous of corporate governance has yet to occur. Present practice is still rooted in a nineteenth century legal concept that is totally inadequate in the emerging global business environment. Present theory is even less capable of explaining coherently the way that modern business is governed. . . . Unfortunately, the most likely driver of further rigorous development in corporate governance is likely to be the next round of alleged board level excesses and corporate collapses, whatever the causes."

Table 4-2 Corporate Governance Theories

Shareholder Theory	Stakeholder Theory
USA and UK	EU and Japan
"Short-term" objectives	"Long-term" objectives
Market-based economies	Relationship-based economies
Focus on shareholders	Focus on stakeholders
Agency theory	Stewardship theory
Outsider systems	Insider systems
Dispersed ownership	Holdings
Stock market	Banks and equity market
Innovation, cost competition	Incremental changes and quality
No external intervention	External regulation
Market is the natural leveling force	Market alone is incapable of correcting disparities and inequalities

Both theories have highs and lows when comparing their competitive advantages and their control and enforcement particularities. The shareholder theory promotes innovation and cost competition, demanding no external intervention in corporate internal affairs, as the market is seen as the natural leveling force; whereas the stakeholder theory eases incremental changes and quality competition, claiming a responsible, participative, and inclusive handling of internal affairs, requiring external regulation because the market alone is perceived as being incapable of correcting disparities and inequalities.

Many voices have proclaimed the superiority of the stakeholder theory, such as De Graaf and Herkströter (2007, p. 179), who said that "the company must not only be accountable to the shareholders. . . . The wider social interest must mainly be protected by (governmental) regulations." They spoke about an interactive and collaborative network-oriented stakeholder perspective of the company. However, when Siemens pleaded guilty to bribery and corruption charges and agreed to pay fines in excess of US$ 1.5 billion, it showed that this is not a bullet-proof theory, and that most of its advantages, such as generation of goodwill, may be barely rhetoric. *The Economist* (2008) titled it "The stench of bribery at Siemens signals a wider rot in Europe" (see Case 4-1).

In 2008, the spread of the financial crisis (which first was associated exclusively with the subprime mortgage industry in the United States) across industries and borders demonstrated that the present models of corporate governance are vulnerable, with plenty of leeway and discretionary power in the hands of managers, who are often beyond practical enforcement.

Case 4-1 Bribery at Siemens

In 2008 Siemens, one of the largest European engineering firms, pleaded guilty to global corruption and bribery, agreeing to pay fines in excess of US$ 1.5 billion.

Siemens' eagerness to win telecom contracts led it to establish "cash desks" from which to withdraw money to bribe foreign officials and politicians. Managers approved their own withdrawals, no questions were asked, and no documents were requested.

Imaginative accounting enabled claiming tax deductions for bribes at least until 1999. Without realizing the inappropriateness of their acts, employees moved money in cash instead of bank transfers, as smugglers. Eventually, business consultants from outside Siemens were used for this scam.

There was a detachment between local and foreign ethics; in some regions or countries perceived as ethically weaker, a different ethical standard was used than was used at home—in this case, Germany.

At that time the laws of most of the European countries were lax regarding foreign bribery. Since then some countries, for example, France and Germany, have made steady improvements.

Questions

1. How do you believe Siemens officials (or anyone else) should proceed if a foreign local official or politician requests money or other favors in order to facilitate or grant a contract in their respective country?
2. Is bribery acceptable if other companies are also bribing in that country?
3. Is bribery acceptable if the company will go out of business without the contract?
4. Do you believe that we should be consistent, with only one set of rules everywhere in the world, or should we adapt to local customs?

As corporations have concentrated wealth and power, concerns have developed to the point of describing corporate power as a growing and insidious phenomenon. The public mindset regarding powerful individuals and organizations is often one of suspicion, fueled by an atmosphere of mistrust within a society that has witnessed evildoing by politicians,* clergy,† athletes,‡ and show business icons§ on a daily basis. A sustained series of revelations of malfeasance related to those who previously were regarded as trusted individuals and leading corporations has seriously undermined public confidence in their probity. Differentiating between veracious practices and window dressing is not easy, as in some organizations disclosure is used primarily in response to public pressures, attempting to alter perceptions without altering the facts.

"Bigger is better" showed that the older corporate governance paradigms were unlikely to contribute to the advance of society, as progress could be limited with the use of corporate power in pursuing of financial gains instead of using that power to explore new and unconventional behaviors that could benefit individual consumers and society at large.

A section of literature on corporate governance has a Darwinian view of organizations, associating appropriate governance structures with a higher chance of survival in competitive markets. The ability of governance structures to adapt to changes in the marketplace is associated with efficient organizational forms, which are prone to survival, while those which fail to adapt face extinction. Consider the fate of big companies such as Nortel and Pan Am, and the prospects of IBM.

Any comprehensive review of the contemporary ongoing nature of corporate governance should address the convergence between legal (compliance-driven) and moral (accountability-driven) liabilities. By doing so, the concept of corporate governance can be related to economic and social goals such as economic efficiency and stakeholder welfare, including the following factors:

- *Business ethics and principles (legitimacy):* ethics, trust, sustainability, fair and responsible remuneration (distribution of wealth), moral purpose of the corporation, leadership and service, diversity and inclusion, market position, reputation, social and environmental interests

---

* For example, Dominique Strauss Kahn (2012), General David Petraeus (2012), Bill Clinton (1995–1997), Mark Sanford (2009), Jesse Jackson, Jr. (2013), Rodrigo Rato (2015), Congressman Christopher Lee (2011), and Congressman Anthony Weiner (2011).

† For example, sex abuse scandals, including the Archbishop of Edinburgh, Cardinal Keith O'Brien (2013).

‡ For example, Tiger Woods (2010), Jerry Sandusky (2012), and Lance Armstrong (1999–2005).

§ For example, Martha Stewart (2010) and Wesley Snipes (2010).

- *Processes and procedures (legality):* corporate structure, objective definition, strategy selection and implementation, management processes and
  policies, laws, traditions and institutions, means to attain corporate objectives, system of direction and control (authority and accountability), roles
  and responsibilities, innovation and learning, access to capital
- *Behavior and practices (participation):* relationships among corporate
  management, the board, shareholders, governmental authorities, and
  other stakeholders; resource allocation, global workforce, employee
  motivation, communication and information disclosure (transparency),
  shareholder and stakeholder rights, expectation settings, performance
  monitoring and optimization, risk management, decision making,
  responsible business practices

4.2 Project Governance

*[No] governance project is complete unless it includes a concern for the
ethics of the proposed actions and a concern for the way that recipients of the
proposed actions will be treated* (Van Gigch, 2008, p. 151).

Today, projects struggle for success in a world with blurred borders. Project
managers recurrently ask themselves about the right structure to help their projects perform, optimize time and resources, and deliver quality and excellence
in a timely fashion.

While corporate governance structures have changed over time, the earliest references to project management, as it is known today, come from the late
19th century, when large-scale projects (such as railroads) demanded a change
in organizational attitude and new organizational structures. It was not until
after World War II that new tools and diagrams were introduced that provided
managers with the means to plan and control complex projects, triggering the
development of common definitions, procedures, and practices.

One of the first challenges was understanding what project management is
and what could be expected from it. Is it a philosophy, a set of tools, processes
and techniques, an art, or a discipline?

Before the 1980s, limited attention was given to the influence of the
organizational environment. The focus was mainly on project team management, leadership, stakeholders' relations, and technical skills.

In 1981, the Project Management Institute (PMI) decided to support the
profession of Project Manager by developing the necessary concepts and procedures in three areas: ethics, standards, and accreditation. By 1983, a report
was generated including a Code of Ethics, a Standards Baseline consisting of six

knowledge areas (Scope Management, Cost Management, Time Management, Quality Management, Human Resources Management, and Communications Management), and guidelines for accreditation and certification. This was the first significant attempt to integrate all the disperse pieces into a unified body (Wideman, 2014).

In 1987, the publication of the PMI's *Project Management Body of Knowledge (PMBOK)* (Project Management Institute, 1987) added the "project management framework" to cover the relationships between general management and project management, as well as the project and its external environment. It also added two new knowledge areas (Risk Management and Contract/Procurement Management). These areas are closely related to ethical issues. The first area considers both the risks during the project as well as the risks associated with the outcome of the project. The second area is to guide, assure, and control that the ethical principles and the economic and regulatory aspects are fulfilled.

By 1996, since the complexity of project management was clear, the PMI published *A Guide to the Project Management Body of Knowledge (PMBOK® Guide)* (Project Management Institute, 1996), changing the title to emphasize that this document was not the Project Management Body of Knowledge but merely a guide. It identified five essentials of project management: a common vocabulary, teamwork, situational application of the management elements, a gated project life cycle, and executive support.

Some people found a shift in focus from the role of project manager as a single great orchestrator of resources and integrative responsibility to one focused on sharing integrative responsibilities among different levels, recognition of the importance of team planning, and also a predictive aspect added to the planning and control functions.

By 2008, the PMI published the Fourth Edition of *A Guide to the Project Management Body of Knowledge (PMBOK® Guide)* (Project Management Institute, 2008) including for the first time a definition of Enterprise Environmental Factors (among them the organizational culture and structure). It recognized that these Enterprise Environmental Factors may constrain project management options, which would influence the project's outcome. The old project management perspective that focused on insider factors of the project thus gave way to a broader one linked to corporate governance.

In 2013, the PMI published the Fifth Edition of *A Guide to the Project Management Body of Knowledge (PMBOK® Guide)* (Project Management Institute, 2013) with the addition of a tenth knowledge area—Project Stakeholder Management—which recognizes the capital importance of engaging project stakeholders in key activities and decisions.

The literature on project governance is diverse and addresses the governance issue from various points of view. In fact, specific literature on project governance was scarce before the last few years, when authors as the PMI (2006), Renz (2007), Garland (2009), Müller (2009), and Crawford (2008) started to look specifically into this subject.

While Renz (2007) introduced project governance as the middle layer between corporate governance and project management, Garland (2009) recognized project governance as a critical success factor for the delivery of projects, Müller (2009) supports that project governance coexist within the corporate governance framework, and Crawford (2008) saw project governance as an enabler of the delivery capability. Others see project governance as the bottom level in governance. As such, it takes policies, processes, and the definition of roles and responsibilities in a top-down fashion, cascading from the corporate level, as an extension of corporate governance. While it is mostly described in terms of the processes necessary for a successful project, its aim is to contribute to corporate strategy through predictable and consistent delivery of projects.

Taketomi (2009) recognized that "governance is necessary to achieve project success, accomplish its original mission, and bring forth the expected return to the project owner."

Some project governance–related definitions are given in Table 4-3.

Although there is almost a general consensus about the aspects covered by project governance, especially when referring to those related to project management functions already depicted in the PMBOK® Guide (Project Management Institute, 2013), some other aspects (such as ethics) are diminished or even completely neglected in the contemporary bibliography.

The reports issued by the Organization for Economic Co-operation and Development (OECD) consistently show a project's high failure rate in terms of budget, functionality, and timeliness. People are asking whether the former project management conceptions hold the line with present-day complexities and to what extent. While mainstream researchers focused on developing new techniques, standardization, and certifications in project management skills, few thought about the project governance role in the observed project's high failure rate.

The World Bank (2006, p. 1) observed that "researchers estimate that when governance is improved by one standard deviation, incomes rise about three-fold in the long run, and infant mortality declines by two-thirds." If improvements in governance have proved to have a positive impact for countries, we could ask: What should be the impact of improving project governance on projects' success?

In contrast to the theories of corporate governance, which are mutually exclusive such that a corporation can apply one (shareholder) or another (stakeholder), in project governance more than one theory can be applied serially. At the beginning, in the early phases of the project (precontract), it might be the

Table 4-3 Project Governance

Alvarez-Dionisi (2008, pp. 56–57)	*Project governance:* "This is the process-driven system that allows management, shareholders, the board of directors, and other stakeholders to have timely, relevant, reliable, and transparent information on all enterprise investments made via projects, programs, and portfolios. Project governance is a subset of corporate governance by which projects, programs, and portfolios are directed and controlled in order to implement the organization's strategy. The executive management and board of directors are accountable for project governance."
Klakegg et al. (2008, p. S29)	*Governance of projects:* "Concerns those areas of corporate governance (public or corporate) that are specifically related to project activities. It consists of formal and informal arrangements by which decisions about projects are made and carried out. Good governance of projects ensures that relevant, sustainable projects and alternatives are chosen, delivered efficiently, and can be cancelled when appropriate."
Müller (2009, p. 9)	*Governance of projects:* "The model by which projects, programs, and portfolios will be governed, including the means of prioritizing scarce resources among projects competing for those resources."
Project Management Institute (2013, p. 34)	*Project governance:* "Provides a comprehensive, consistent method of controlling the project and ensuring its success."
Raterman (2003, p. 1)	*Project governance:* "Ensures a project is completed according to plan and that its ultimate business objectives or benefits are delivered."
Renz (2007, p. 19)	*Project governance:* Is a process-oriented system by which projects are strategically directed, integratively managed, and holistically controlled in an entrepreneurial and ethically reflected way, appropriate to the singular, time-wise limited, interdisciplinary, and complex context of projects.
Turner (2009, p. 61)	*Governance of a project:* "This involves a set of relationships between the project's management, its sponsor (or executive board), its owner, and other stakeholders. It provides the structure through which the objectives of the project are set, and the means of attaining those objectives and monitoring performance are determined."

transaction cost economics (TCE) theory, which is used to explain the different governance structures that are adequate for different types of contracts. In later phases of the project (postcontract), it might be the agent theory, which is used to explain the information imbalances between the sponsor (principal) and the project manager (agent).

There is no universal project governance model. Project governance should fit the context of the organization strategy and the environment, setting a framework that deals with accountabilities, strategic alignment, decisions, roles, management, control, and ethics. Project governance aims for transparent, repeatable, and scalable principles, structures, and processes, which require steady review to maintain effectiveness.

Müller (2009, pp. 10–12) described four paradigms of project governance as the result of combination of the behavior (rules) or outcome (principles) orientation of the organization and the shareholder or stakeholder theory of corporate governance endorsed. He asserted: "The choice of governance paradigm impacts the breadth and depth with which governance of projects and project management is implemented in an organization."

One key differentiator of this modern vision of project governance is that it recognizes the classic top-down approach developed from the outcome orientation, while it introduces a bottom-up approach developed from the behavior orientation applying project management methodologies.

As projects are unique and there are many different types of projects, project governance necessarily is a dynamic matter, meaning that the project governance paradigm also depends on the type of project. This also means that there may be different ways of governing projects even within the same organization or community. Table 4-4 accounts for some common types of projects in a for-profit organization.

Two lines of thought can be recognized in project governance. One endorses the application of rules, and the second advocates for the application of principles. While rules may be easier to prescribe and follow than principles, making a clear distinction between acceptable and unacceptable behavior, they also reduce the project manager's ability to act discretionally. In practice, rules were demonstrated to be well suited for static situations and bureaucratic top-down structures. They are good references to deal with the status quo, but they may be unable to deal with change such as unpredicted technological breakthroughs and new transactions that may fall outside the scope of rules. Even when clear rules are enforced, daily experience demonstrates that there is always a way to circumvent their underlying purpose.

In contrast, in a self-regulated way, principles allow for the determination of which standards are acceptable for every situation, under the umbrella of the wider scope of project integrity. They can avoid flaws related to attachment to impractical or outmoded rules.

Table 4-4 Project Types

Risk Orders
The project starts with no Purchase Order received from the customer.
An audit trail of authorization from senior management must authorize its cost.
The revenue will come from the customer.
The project type will change to Customer Billable once the Purchase Order is received.
Early Ramp-up
An audit trail of authorization from senior management must authorize its cost.
The main project is constrained by time, so Early Ramp-up work is required to meet deadlines.
The revenue will come from the customer.
Customer Billable
Purchase Order has been received from the customer.
The revenue will come from the customer.
Internal Recharge/Internal Business Development
Project has no revenue.
Project cost is charged internally.
Trial
The customer has no commitment to pay.
An audit trail of authorization from senior management must authorize its cost.
Project cost is charged internally.
Pre-Sales
The project starts with no Purchase Order received from the customer.
An audit trail of authorization from senior management must authorize its cost.

While projects tend to be governed at all levels by applying the same line of thought (rules or principles), this should be considered a mixed approach (where some positions in the project are best suited to rules, while others are best suited for principles), instead of making an absolute judgment of value and letting one or another line of thought affect all the positions within the project. Positions concerned with the delivery of the project scope and with the wider organizational impact of the project seem to be a good match for rules, while other positions, concerned with the achievement of high-level project outcomes and with the development of strategic capabilities, seem to be a better fit for principles.

The extent of project governance enforcement depends on the regulatory system, which is directly related to its credibility. A regulatory system does not refer exclusively to government intervention; it also considers corporate policies and project self-regulation. In mature societies, this system is constructed from social uses and values and from the proactive intervention of social bodies and individuals acting as concerned stakeholders. Enforcement dissuades dubious actors, building common ethical standards. The degree of enforcement and chances for innovation, change, risk taking, and progress should, however, be considered as being in a delicate equilibrium. Too much enforcement may dampen creativity, too little may foster fraud and deception.

The Handbook of Project-Based Management (Turner, 2009) identifies three levels of governance within a project-based organization: Executive Board Level, for key and large project investments; Context Level, including the developing of the program and portfolio management infrastructure to link projects to corporate strategy, and the capability development within the organization to deliver the project successfully; and Individual Project Level, concerning the governance structure and roles of the individual projects.

A weak support for ethics at the project level has been observed, as this support is mainly considered a top-down approach under which ethics belongs to corporate governance and spills down to projects. It may work for long-lasting projects with strong bonds to the organization, but this approach has proved to be misguiding in contemporary project work that involves a large number of heterogeneous contractors and third parties for which corporate ethics spillage is dubious. A singular project culture is rarely built, as the different cultures of its members are difficult to mix during the project's time span. In many cases, no ethical contagion should be expected from the top-level executives; project managers should have their own ethics culture that is shared and enforced among all the stakeholders.

Renz (2007) mentioned *integrity management* as a process model based on the combination of discourse ethics and recognition ethics. Müller (2009) referred to the use of legitimacy by the Institutional Theory to stress conformance with society as well as stakeholders' values and expectations. Of the six project governance models described by Alvarez-Dionisi and Turner (2012), only three have some references to ethical boundaries. For those who have a process-oriented view, project governance is a system for structuring, running, and controlling a project. Others see it as a commitment to values and ethical conduct in the management of a project.

Effective project governance is a fundamental requirement for project success. Some commonly agreed-on principles of project governance are regard for the stakeholders, integrity and ethical conduct, transparency and opportunity in the disclosure of information, and executive control.

In spite of the common agreement about these principles, we found that in many projects some of them are not followed (see Case 4-2). We also found that the frequency and scale of the violations are not always directly related to its impact on the project performance. There are instances where a single failure to adhere to one of these principles is enough to ruin the whole project, while in other instances a steady pattern of misconduct cannot derail the project.

Case 4-2 Project Celebration

In a Scandinavian telecom company there was a tradition of caring and rewarding the project teams with events to celebrate the achievement of milestones during the project life cycle. Those celebrations usually consisted of a dinner followed by the visit to a local pub, which the project manager paid for with a company expense card, charging the cost to the project.

During a difficult moment for the telecom industry, it happened that the company changed some policies. Restrictions were imposed on travel and expenses, and project celebrations were explicitly banned.

Project SCM-10 had one milestone celebration event already planned and communicated to the project team members when the new rules were announced. The project was also running below its cost baseline, leaving room to accommodate the expenses of the celebration. Eric, the project manager, decided to proceed with the event and asked Stefan, an external contractor, to pay the bill, invoicing the amount as working hours.

Stefan was not aware of the new company policies, but he knew that invoicing expenses as working hours was illegal, although this practice was not uncommon in the industry. As Eric was a resourceful leader who always found ways to keep his team motivated, Stefan felt his assignment might be threatened if he refused Eric's request.

Eric believed that his solution was the best for the company because motivated team members do perform better. He also considered that higher management was wrong by changing the policies and thus interfering with his handling of the project budget.

This case illustrates how changes in corporate governance and policies can challenge ethical behaviors at the project level. As change is not always welcome, new policies need to be explained, discussed, rationalized, and then enforced. Simple enunciation is not enough to guarantee adherence.

Questions

1. Did Stefan do anything wrong? Did he have any other alternative?
2. Did Eric do anything wrong?
3. As a project manager, what was Eric's main responsibility: keeping his team motivated or complying with new corporate policies?
4. Should Eric have escalated this situation, explaining that the event was already planned and communicated, or did the magnitude of this small expense not justify an escalation? After all, as project manager he was making decisions of more value on a daily basis.
5. Does this situation even constitute an ethical issue?

The quality of project governance is determined by both external and internal forces. Financial markets, legislation, and the international organizational environment count among the external forces, while implementation of policies and processes and leadership are among the internal forces. Although to a large extent external forces cannot be controlled by the project, internal forces, in contrast, provide the elements and opportunity for differentiation.

4.3 Summary

This book is looking primarily into the legitimacy layer (ethics and principles) and the promotion of ethical and responsible decision making. Its main view is aligned with the principle that projects should not only comply with their legal obligations but should also consider the expectations of their stakeholders, including shareholders, employees, customers, suppliers, and the broader community in which they operate.

It can be said that any comprehensive review of the contemporary ongoing nature of project governance should compound all the definitions previously discussed. By doing so, the concept of project governance will then be related with the following factors:

- *Business ethics (legitimacy):* fairness, rules, and principles
- *Processes and procedures (legality):* project structure, corporate strategy, processes and policies means to attain project objectives, system of direction and control (authority and accountability), project risk management,

roles and responsibilities, predictable and consistent delivery, project scope (functionality), budget, efficiency, execution

- *Behavior and practices (participation):* Relationship among project managers, sponsors, and other stakeholders, resource prioritization, incentives, disclosure and transparency on material matters of the project, performance monitoring, decision making, timeliness

Chapter 5

Small Sins Allowed and the Line of Impunity

This book attempts to uncover constructs, variables, and relationships involving ethics and governance within the project management field. In doing so, the historical perspective was reviewed, exploring the views of ethics "including ethical decision making models, economization, double standards, codes of ethics, and ethical dilemmas" (Lopez & Medina, 2015, p. 588), as well as context (perceptions, motivation, and culture), and governance (corporate and project).

We have stated that ethics is related to the influences received at home, school, church, and from the legal system. It lies underneath the law, deep inside the individual and social consciousness, as a complex construct of universals (the portion of ethics norms widely accepted) and local beliefs (principles confined to a particular group), which include integrity, justice, competence, and utility. "There is a general agreement that legality cannot involve all perceptions of morality" (Lopez, 2015, p. 142).

"The ethical decision-making process is an issue-contingent matter, meaning that the characteristics of the moral issue (moral intensity) are important determinants in this process" (Lopez, 2015, p. 41). As many other variables of the context affect the perception of the ethical climate, deviance should not be referenced to an absolute system of moral standards. In this sense, since the meaning of morality seems to shift according to economic and cultural values, social structures exert pressure on certain individuals to engage in nonconformist conduct.

Workplace deviance differs from ethics. Contextual conditions such as an employee's adhesion to project structure, characteristics of the rules, and the individual normative cultural perspective determine which rules are more likely to be violated.

Other related concepts frequently found include:

1. References to the perception of certain thresholds or levels of tolerance toward rule violations, such as the level to which managers believe that ethically suspicious acts are justifiable, behavior that falls short of the moral standards, and the mention of basic levels of ethical behavior. It is also frequent to find mentions related to behaviors that can be overlooked or ignored because of their low moral intensity or because of subjective morality, such as minor infractions of social norms, situations of mild abuse, and the "never mind the rules" disposition.
2. References to the perception that factors such as power distance, status, privileges, social stratification, position, double standards, executive accountability, elites, no concern about penalization for ethical violations, power as the source of corruption, making one's own rules, and being above the rules can be related to the proclivity toward ethical violations through deviant behaviors.

Lopez (2015, p. 180) conducted a survey in order to capture the perception of project managers regarding proclivity toward deviant behavior and to test the hypothesis that some deviant conduct increases with hierarchy.

As a first step, four hierarchy levels were defined: freshman, sophomore, junior, and senior, which could apply to any kind of organization, such as industry, the military, clergy, academics, politics, etc. Each of these levels simply attempted to reflect the seniority of their members, which could be linked with their behaviors. Then, only five subjects or areas where deviant behaviors could occur at a project level were chosen: breaking the law, project property abuse, improper moral conduct, unsocial behaviors, and discrimination.

The respondents' perception was that at a freshman level there is a proclivity to agree with deviant behaviors that exceeds all other levels for any considered subject. Conducts related to breaking the law and improper moral behaviors topped the deviant ranking. Conducts related to the abuse of project property showed disregard toward project assets. On the other hand, conducts related to embracing unsocial behaviors and discrimination showed the lowest proclivity.

Two comments were collected from the freshman level:

1. What an entry level person believes is correct, by the time he or she reaches a senior level the person knows is not acceptable.
2. Someone new will more than likely follow the rules.

These two statements show the difference in criteria among respondents. While one points to inexperience as the source for deviant behaviors, which will eventually disappear with seniority, the other believes that inexperienced employees will follow the rules, suggesting that this conduct may fade away with time.

When considering the sophomore level, the only subject that was still clearly skewed toward a deviant behavior was breaking the law, while all the other subjects decreased considerably. This suggests the perception that sophomores have a lower proclivity than freshmen toward deviant conduct (see Figure 5-1).

Three comments were collected from the sophomore level:

1. Letting people feel that "having more power" in the team could result in better output.
2. Some rules start to be more relaxed.
3. Over time, initial hesitation to do certain "not completely acceptable behaviors" will diminish as the individual's status within the group is established, and the people learn the inner workings of the project.

While the first statement shows the belief that empowered individuals could do better, the other two relate to relaxing rules with hierarchy and to the time and status required to accommodate deviant conduct to the project's expectations.

The junior level showed that all subjects' proclivity toward deviant conduct decreased; in particular, "breaking the law" decreased significantly.

Six comments were collected from the junior level:

1. The more seniority one has, the professionalism expectations should increase.
2. The "higher" one's position (ranking) in one project is, the closer one feels to the "Project," which makes the person feel more responsible for whatever happens within the project.
3. It is nearly impossible to be completely unbiased after moving through these positions of power.
4. Some rules are relaxed while one is learning not to do other things.
5. Usually, as people develop their careers, certain "privileges" are expected to be granted. Sometimes such privileges arise only by changing a previous attitude toward one or more project rules.
6. The higher up in various positions, the more likely the individual believes the rules are for others.

As for the previous levels, there are two kinds of statements. The first two refer to professionalism and responsibility gained at higher levels, while the

	Freshman			Sophomore			Junior			Senior		
	Agree	Disagree	Deviant Ranking	Agree	Disagree	Deviant Ranking	Agree	Disagree	Deviant Ranking	Agree	Disagree	Deviant Ranking
Breaking the Law	57%	26%	1	49%	27%	1	32%	42%	1	36%	51%	1
Unsocial behavior	33%	54%	4	20%	56%	4	14%	74%	4	19%	73%	4
Discrimination	18%	53%	5	15%	54%	5	11%	72%	5	12%	75%	5
Project property abuse	43%	48%	3	35%	53%	3	27%	57%	2	27%	64%	2
Improper moral conduct	53%	30%	2	36%	37%	2	22%	62%	3	20%	70%	3

Figure 5-1 (Top)

Figure 5-1   Perception of Deviant Behavior Proclivity by Subject

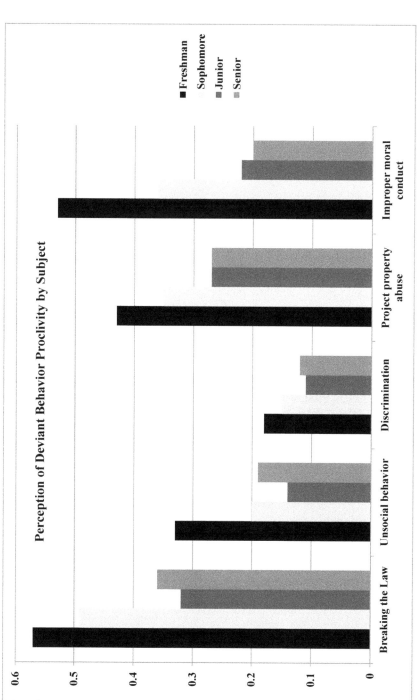

Figure 5-1 (Bottom)

other four refer to relaxation of rules, privileges, and to the belief that rules are meant for others.

The senior level showed a similar deviant ranking as the junior level, with breaking the law on top. It is remarkable that four of the five rates of agreement with deviant conduct increased from the previous level. This shows the perception that the senior level is weaker than the junior level regarding deviant conduct. In particular, the proclivity toward breaking the law is 32 percent, meaning that one out of three seniors is supposed to embrace this kind of conduct.

Four comments were collected from the senior level:

1. They are more likely to want to make their own rules.
2. I think the dating thing would seem fine to the new person (who does not know the rules) and the person in charge (who is above the rules).
3. These people would try to follow the rules to get promoted and then think they are above the rules after getting the promotion.
4. Power corrupts . . . absolute power corrupts absolutely!

In this case the comments are related to being above the rules, making their own rules, and the relation between power and corruption.

When considering the answers by subject, it is noticed that conduct related to breaking the law tops the deviant ranking at all levels. Improper moral conduct is seen as the second deviant offender for the freshman and sophomore levels, and the third deviant offender at the junior and senior levels. Company property abuse is seen as the third deviant offender at the freshman and sophomore levels, and the second deviant offender at the junior and senior levels. Embracing unsocial behaviors and discrimination are at the bottom of the deviant ranking for all levels.

Some other comments collected were:

1. As a women working in a male-dominated industry and moving up the ranks, the further I have gotten and the more contact I have with high-ranking males, the weirder, and comments are more inappropriate.
2. I do not think the person's position matters if they have the mindset they can do no wrong, and rules do not apply to them.

The first statement, which comes from a real-life female experience (not a perception), refers to a gradual decline in ethical standards with hierarchy. The second comment is related to the unconscious transgressions ("the individuals do not even realize that they are making an inappropriate decision, as they fall prey to ethical fading or to other cognitive biases") (Manzoni, 2012, p. 2).

When adding together the five subjects' perceptions of deviant behavior propensity for each level, it was found that the higher proclivity was perceived for the freshman level, with a gradual decrease through the sophomore level, reaching the lowest perceived proclivity at the junior level (see Figure 5-2). What these results suggest is that a reversion in this trend is perceived at the senior level. It is associated with the "perception that deviant conducts are more common at senior level than at junior level" (Lopez, 2015, p. 198), indicating that the learning curve that has been reducing the deviant behavior proclivity so far seems to be influenced by some new variable over the junior level, and this influence is negative.

While 25 percent of these comments refer to inexperience as the source of deviant conduct, the other 75 percent refers to power as the source of deviant conduct. In particular, all the comments related to the senior level refer to power as the source of corruption, making their own rules, and being above the rules.

This study revealed two lines of thought: those who associate deviant behaviors with inexperience, believing that time and hierarchy should help to improve these weaknesses; and those who associate deviant conduct with the power granted at higher levels of the hierarchy.

Freshman	Sophomore	Junior	Senior
204.00	155.00	106.00	114.00

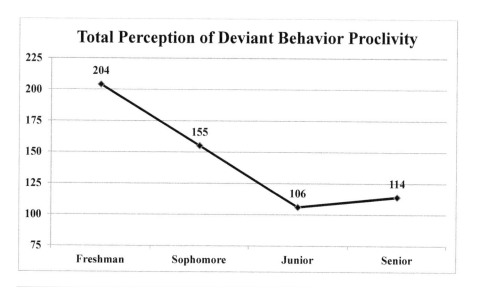

Figure 5-2   Total Perception of Deviant Behavior Proclivity

In line with the first group, Piper et al. (1993) discussed whether ethics can be taught or not, finding that lack of experience and lack of comprehension hinder freshmen. In regard to the latter group, Chetkovich and Kirp (2001) found a bias toward heroic protagonists on top of the hierarchy.

One of the conclusions of the study was that some deviant conduct increases with level in the hierarchy. In addition, two definitions were developed: *Small Sins Allowed* and *Line of Impunity*. Those are described in the following sections.

5.1 Small Sins Allowed

> Small Sins Allowed *is defined as a subjective mental model that establishes the level of a certain behavior above which adherence to ethical standards is expected. It can also be thought of as the ethical tolerance level that splits any dimension into two domains; above this level there are ethical standards to comply with and abide by, whereas below it there are no ethical concerns. This level represents a limit to the freedom of deviating deliberately from customary ethical behaviors and normally applicable rules and practices. No evil in the sense of morally objectionable behavior such as immorality, iniquity, malevolence, or viciousness is perceived below this level, where actions may be perceived by some as insignificant, minimized, understated, or belonging to the domain of permissiveness. Enforcement and punishment are relevant only above this level, where compliance with standards and rules is required* (Lopez, 2015, p. 199).

Fraedrich (1992, p. 13) was the first to use the term sins in association with unethical behavior in his article, "Sins and Signals of Unethical Behavior," although his concepts were not related to the definition of *Small Sins Allowed*.

Furthermore, Bitektine (2008, pp. 73–76) found that:

> *Social norms often act as threshold values or behavioral boundaries, separating what is considered socially acceptable from what is not. The presence of some critical value of conformance with stakeholder expectations . . . is reflected in the concept of a legitimacy threshold.*

In addition, Baucus et al. (2008, p. 104) said that "law represents society's minimum standard of acceptable behavior; to fall below that standard brings sanctions. Regrettably, behavior that is legal may not be broadly discerned as ethical." The concept of a *legitimacy threshold* is correlated to a certain extent to *Small Sins Allowed,* as it creates a boundary between socially accepted behaviors (good) and socially rejected behaviors (bad); however, the concepts are not the same, as *Small Sins Allowed* creates a boundary between the fields of permissiveness and compliance, where permissiveness does not necessarily mean socially accepted behavior (good), but socially tolerated behavior instead (bad). In addition, *Small Sins Allowed* differs from law in the sense that law establishes a rigid and sometimes outdated legal standard, while *Small Sins Allowed* refers to updated socially accepted ethical standards. This way, some behaviors outside the minimal standard defined by law can still be socially accepted, while other behaviors, although complying with the law, are deemed unethical by society at large.

When looking at workplace deviance (violation of organizational norms), it differs from ethics (right or wrong behaviors in terms of justice, law, and morality) in that it arises as the result of a lack of motivation to conform to normative social expectations or as the result of having the motivation to violate those expectations. According to Robinson and Bennet (1995, p. 565), deviant workplace behavior and ethics vary within two dimensions: magnitude (minor versus serious), and scope (interpersonal versus organizational). The distinction between interpersonal and organizational dimensions is aligned with the existence of more than one set of ethical values.

When Robinson and Bennett (1995, p. 557) defined workplace deviance and employee deviance, they focused on "violations of norms that threaten the well-being of an organization, excluding minor infractions of social norms, such as wearing a suit of the wrong style to the office, that are not usually or directly harmful to most organizations." By doing that, they explicitly recognized the existence of the *Small Sins Allowed*. They went even further by enumerating a series of organizational and interpersonal workplace deviant behaviors which lie under the categories of minor and serious (1995, p. 565, Fig. 2). Their list of minor deviant behaviors constitutes a clear count of what we defined as *Small Sins Allowed* "(leaving early, taking excessive breaks, working slowly, wasting resources, showing favoritism, gossiping about co-workers, blaming co-workers), while the list of serious deviant behaviors considers plain felonies (sabotage, accepting kickbacks, lying, stealing, sexual harassment, verbal abuse, endangering co-workers)" (Lopez, 2015, p. 202).

As introduced in Chapter 2, "the ethical decision-making models recognize that the context (organizational, cultural, economic, and social) influences the recognition of a moral issue, which is the first milestone in ethical decision making. Only after the moral issue is recognized, a moral judgment can be made. Then, with the intervention of factors, such as opportunity and individual moderators, the models advance to the establishment of a moral intent to finally engage in a moral behavior (ethics)" (Lopez, 2015, p. 41).

If during the first stage of a decision-making process (recognition of the moral issue), the moral issue is not recognized, this is an indication that this particular issue falls into the category of *Small Sins Allowed*. It will lead in the second stage of the decision-making process (moral judgment) to an unconscious transgression (the individual does not even realize that he or she is making an inappropriate decision). On the other hand, if the moral issue is recognized and, during the moral judgment, the individual decides to pursue it anyway (conscious transgression) because it is deemed too small for serious consideration (low moral intensity), this issue also falls into the category of *Small Sins Allowed*. For one reason or another, *Small Sins Allowed* are excluded from the moral intent and consequently from the moral behavior.

It is widely recognized that a basic level of ethical behavior is required for society's economic and social progress, which is adversely affected by project misconduct. A four-level hierarchy of ethical behaviors can be conceived, where a theoretical level reflects the highest potential for good and the spirit of morality. Below it, a practical level involves extreme diligence toward moral behavior. Following, an attainable level reflects behaviors viewed as moral by society. And finally, a basic level refers to the bare letter of the law, the minimal acceptable behaviors.

Small Sins Allowed fall outside this hierarchy of ethical behaviors, but they are still tolerated, representing a buffer zone between accepted and reprobated behaviors.

Small Sins Allowed can be linked to subjective morality in the sense that they are *"defined as a subjective mental model that establishes the level of certain behaviors"* (Lopez & Medina, 2015, p. 588). They are also related to uncertainty avoidance (formalization, standardization, and ritualization), the meaning of time (precision and punctuality), showing or hiding emotions, and tolerance for deviant ideas and behaviors (Lopez, 2015, p. 202).

Social structure generates the circumstances in which violation of social codes constitutes a normal response (Merton, 1938). "The exposure to postures favorable to vicious acts leads to the infringement of rules (differential association). 'Everybody else does it' is a rationalization for making a poor ethical choice" (Lopez, 2015, p. 39). The broken window theory (Levine, 2005) suggests that any small indication that something is amiss and not being taking care of sends signals about a lack of order, leading to lawlessness, and anarchy.

Baucus et al. (2008, p. 103) found that the "never mind the rules" disposition of some employees allows them to "decide key ethical issues on their own. Minimally, that set of issues would include: (a) which rules to break; (b) under what circumstances rules should be broken; (c) how far to go in breaking the rules; and (d) who gets to make or break the rules." They also said (2008, p. 105) that "some rules should never be broken." These findings mean that employees can choose which rules are among the *Small Sins Allowed* and which are not. It also means that there is a context framework (what, when, how long, and who) that may affect differently the rules to be broken (*Small Sins Allowed*).

In behaviors such as theft, embezzlement, sabotage, vandalism, computer fraud, absenteeism, situations of mild abuse, "managers' willingness to justify ethically suspect behaviors" (Cullen et al., 2004, p. 415), among others, the magnitude of deviance (moral intensity) is a factor in determining when it falls into *Small Sins Allowed* and when into felonies.

5.2 Focus Groups

As groups best enable the exploration of surprise information and new ideas, the definition of *Small Sins Allowed* was discussed by a focus group. "That discussion provided the opportunity for a flexible, free-flowing format. Spontaneous responses were expected to reflect the genuine opinions, ideas, and feelings of the group members about the topic under discussion" (Lopez, 2015, p. 220).

The sessions were structured according to the following scheme. First, during a free association phase, the facilitator asked what words or phrases come to mind when the definitions explored were named. It was seen that their simple mention was basically intuitive, as comments were aligned in some sense with the meaning of both constructs. Second, each definition was defined, and further comments collected now under the light of a clear definition. Third, three citations were mentioned by the facilitator to trigger the discussion. Fourth, a role-play phase was conducted around one or two citations. As a result of the two focus groups sessions, the definitions were evaluated.

5.3 Focus Group Session #1: Small Sins Allowed (Lopez, 2015, pp. 221–227)

Free Association

The focus group section started with the following question:

> *What words or phrases come to mind when you think about Small Sins Allowed?*

Responses

> *Low-risk judgments and complaints.*
> *Employee deviance.*
> *Things that you identify at the beginning of a project that you expect not to do during execution, but you end up doing anyway.*

Low intensity is allowed; high intensity is not.

Certain low risks can be accepted, while higher risks are not.

Senior management can override some procedures.

If the outcome of a risk result is bad for the organization, then it is considered a "sin"; in contrast, if the outcome of taking a risk is positive, no "sin" is associated with that risk.

Success is not a "sin," while failure is.

The second question was

Small Sins Allowed is defined as a subjective mental model that establishes the level of a certain behavior above which adherence to ethical standards is expected. It can also be thought of as the ethical tolerance level that divides any dimension in two domains; above this level there are ethical standards to comply with and abide by, whereas below it there are no ethical concerns. This level represents a limit to the freedom of deviating deliberately from customary ethical behaviors and normally applicable rules and practices. No evil in the sense of morally objectionable behavior such as immorality, iniquity, malevolence, or viciousness is perceived below this level, where actions may be perceived by some as insignificant, minimized, understated, and belonging to the domain of permissiveness. Enforcement and punishment are relevant only above this level, where compliance with standards and rules is required (Lopez, 2015, p. 199).

Responses

In my personal life I cannot take any minimal deviance, even a 50 cent candy bar cannot be stolen, but at the office some stuff is there for your use, and if you take it home it could be OK depending on the value.

It depends on where you are, in which country. In China bribery is very common, while in Canada it is not.

People justify misrepresenting their expense reports or taking office stuff home for their kids because they believe this is not stealing.

Time reports are frequently exaggerated. People think "I worked so hard that I deserve claiming a few extra hours."

Mileage claims are also inflated. People feel they are helping the company by using their own cars, and the mileage fee does not cover all the inconvenience.

Double standards are common, as well as conflict of interest.

Reporting of key performance indicators (KPIs) is skewed because everyone wants their KPIs to be green. So even when a yellow is the right representation, it is overridden and avoided in the reports.

The third question was

Do you believe that institutions depend on a basic level of ethical project behavior?

Responses

A lot of small sins could add up and exceed the basic level of project behavior.
Cultural behaviors override codes of conduct.
Certain levels of dysfunction can always be managed.

The fourth question was

"Workplace deviance (violation of organizational norms) differs from ethics (right or wrong behaviors in terms of justice, law, and morality); it arises as the result of a lack of motivation to conform to normative social expectations or having motivation to violate those expectations" (Robinson & Bennett, 1995, p. 556). *Do you believe that in a certain sense, by differentiating workplace deviance and ethics, this assertion is putting workplace deviance among the Small Sins Allowed, while ethics is above them?*

Responses

They cannot be separated, as they come together.
For knowledge workers you need to consider that they cannot be using their minds 100% of the time, so checking their personal emails, reading the sport news, or taking breaks could be considered as part of their creative process.
If you consider productivity instead of attachment to the office time schedule, it is hard to define a black and white scenario.
Arriving late or leaving early consistently could be compensated by some work done at night from home, so there are many shades of gray to consider.
If you let the small things continue to add up, at some point you will absolutely hit the ethics basics.

The fifth question was

Objective and subjective morality were distinguished by De George (1999), who defined objective morality as the broadly agreed societal moral law, which corresponds with promulgated laws. In contrast, he compared subjective morality to the individual's belief regarding the appropriateness or incorrectness of an action, which corresponds with the concepts of moral sense, scruples, and conscience. Intrinsically, decisions must be based on both objective and subjective moralities. Do you believe that the described subjective morality resulting from an individual's belief can be linked to the Small Sins Allowed, in the sense that they are defined as a subjective mental model that establishes the level of a certain behavior?

Responses

Most people think of morality as objective, although mostly it is a subjective matter.
Morality changes from country to country due to cultural differences.
When I was younger I had a different moral mindset, it seems to change over time.
Even when you apply your subjectivity, there are lines that cannot be crossed.
You should have a basic moral stand for society to function.
We became more tolerant as society became more diversified.
Thinking about "too big to fail" and the "Bill Clinton affair," society became tolerant and accommodative with its standards.

Role-Play

Statement 6: The "managers' deliberative reasoning processes" (Thorne & Saunders, 2002) concerning the level to which they believe that ethically suspicious acts are justifiable, were labeled by Cullen et al. (2004, p. 412) as "manager willingness to justify ethically suspect behaviors."

Responses

I saw a company where bonuses were attached to the "bad debt" structure. Some employees reported collections as delayed before the end of the period and moved them to "bad debt" after the bonus was paid. Their manager knew about this situation but allowed them to proceed as their accounts were highly regarded by the company.
I knew a rude leader that treated his team harshly. Although it was known by the management, he was allowed to continue because his division results were overwhelming. This was frustrating.
At the end of the day it all comes back to results for the company.
The bonus structure has a lot to do with this kind of managerial reasoning.
I don't believe this affects the company's ethical stand, unless there is a real misrepresentation or a falsifying of the numbers.
It could be considered a managerial tool to lead the team.
It generates stress among team members and unfair workload and rewards.

Final Comments

I think small sins are allowed. In my house when the phone rings I tell my daughter to say I am not home.

How much do we really allow?
Training could help people set the bar.
It is really important for leaders to behave ethically and to clearly define what is small and what is big. They should be modeling the system.

5.4 Summary of Small Sins Allowed

The basic level of ethical behavior refers to the bare letter of the law, the minimal acceptable behaviors. As legality does not invariably involve the totality of an issue's perceived morality, some behaviors that are not punished by the law may still not be socially accepted (legitimacy threshold), meaning that even when law does not condemn those behaviors, society does.

Small Sins Allowed are excluded from the moral intent and consequently from the moral behavior. They lie between permissiveness and abidance, where permissiveness does not necessarily mean socially-accepted behavior (legitimacy threshold), but socially tolerated behavior instead (Lopez, 2015, p. 200).

Minor workplace deviant behaviors and minor misconducts (infractions) are among the *Small Sins Allowed* (leaving early, taking excessive breaks, working slowly, wasting resources, showing favoritism, gossiping about co-workers, blaming co-workers, wearing a suit of the wrong style to the office, tardiness, absenteeism, and poor work performance) (Lopez, 2015, p. 202).

Serious deviant behaviors and major misconduct (felonies), such as sabotage, accepting kickbacks, lying, stealing, sexual harassment, verbal abuse, endangering co-workers, and substance abuse, are above tolerance and require punishment (see Figure 5-3).

The focus group discussion, along with the survey and the research on *Small Sins Allowed,* can be summarized as follows.

Subjective morality is recognized as one of the main drivers of the *Small Sins Allowed (most people think of morality as objective, although many times it is a subjective matter; at the end of the day it all comes back to results for the company;*

Ethical Hierarchy	
Serious Workplace Deviance and Misconduct	Felonies
Small Sins Allowed	Socially Tolerated Behavior
Minor Workplace Deviance and Misconduct	
Legitimacy Threshold	Socially Accepted Behavior
Basic Level of Ethical Behavior	Law

Figure 5-3 Ethical Hierarchy

success is not a "sin," while failure is; if the outcome of taking a risk is bad for the organization, then it is considered a "sin"; in contrast, if the outcome of taking a risk is positive, no "sin" is associated with that risk.)

It is associated with:

1. Multiple ethical mindsets *(in my personal life . . . but at the office; it depends on where you are; morality changes from country to country; when I was younger I had a different moral mindset, it seems to change over time; in my house when the phone rings I tell my daughter to say I am not home).*
2. Moral intensity *(depending on the value; there are lines that cannot be crossed; how much do we really allow?)*
3. Cultural differences *(cultural behaviors override codes of conduct. Morality changes . . . due to cultural differences).*
4. Managers' willingness to justify ethically suspect behaviors *(senior management can override some procedures; . . . everybody wants their KPIs to be green. So even when a yellow is the right representation, it is overridden and avoided in the reports; the bonus structure has a lot to do with this kind of managerial reasoning; it could be considered a managerial tool to lead the team; I saw a company where bonuses were attached to the "bad debt" structure. Some employees reported collections as delayed before the end of the period and moved them to "bad debt" after the bonus was paid. Their manager knew about this situation but allowed them to proceed, as their accounts were highly regarded by the company; I knew a rude leader that treated his team harshly. Although it was known by the management, he was allowed to continue because his division results were overwhelming; it generates stress among team members and unfair workload and rewards).*

Other drivers of *Small Sins Allowed* are

1. Uncertainty avoidance *(certain low risks can be accepted while higher risks cannot; a lot of small sins could add up and exceed the basic level of project behavior; certain levels of dysfunction can always be managed).*
2. The meaning of time *(for knowledge workers you need to consider that they cannot be using their minds 100% of the time, so checking their personal emails, reading the sport news, or taking breaks could be considered as part of their creative process; if you consider productivity instead of attachment to the office time schedule, it is hard to define a black and white scenario; arriving late or leaving early consistently could be compensated by some work done at night from home, so there are many shades of gray to consider).*
3. Tolerance for deviant ideas and behaviors *(low intensity is allowed, high intensity is not; double standards are common, as well as conflict of interest;*

we became more tolerant as society became more diversified; thinking about "too big to fail" and the "Bill Clinton affair," society became tolerant and accommodative with its standards).

The exposure to postures favorable to vicious acts (differential association) positively predispose the commission of *Small Sins Allowed*, particularly:

1. "Everybody else does it" *(people justify misrepresenting their expense reports or taking office stuff home for their kids because they believe this is not stealing; time reports are frequently exaggerated. People think "I worked so hard that I deserve claiming a few extra hours"; mileage claims are also inflated. People feel they are helping the company by using their own cars and that the mileage fee does not cover all the inconvenience).*
2. "Never mind the rules" *(employee deviance; things that you identify at the beginning of a project that you expect not to do during execution, but that you end up doing anyway).*

A belief that managers are a strong factor in *Small Sins Allowed* was observed. For this reason, it is important for project managers to behave ethically and to define clearly what is small and what is big. In this sense, training can help in setting the bar.

5.5 The Line of Impunity

The Line of Impunity *refers to the idea that certain ranks or positions in the social hierarchy entitle prerogatives or advantages, and that the power granted at those levels transcends the limits of control or law enforcement (such as a tyranny). Overcoming the Line of Impunity consequently implies a rise in the ethical tolerance level* (Small Sins Allowed), *and hence a detachment from the customary ethical behaviors still demanded for lower ranks* (Lopez, 2015, pp. 205–206).

Power distance is defined as "the extent to which the members of a society accept that power in institutions and organizations is distributed unequally" (Hofstede, 1984, p. 83). It affects the behavior of all members of society. Societies with large power distance easily accept a hierarchical order with a place reserved for everyone; in contrast, societies with small power distance call for explanation of inequalities and strive for equalization. In general, smaller power distance correlates with more developed countries, while larger power distance correlates with less developed countries. There is not inevitably more abuse of power in cultures with large power distance than in those with small power distance. In small-power-distance cultures, status is usually achieved by merit; in large-power-distance cultures, status is often assigned based on ancestry, wealth, rank, or the like. In this case, power differences are read into visible status differences between superiors and subordinates.

While a system of checks and balances against power abuse is common in low-power-distance cultures, in high-power-distance cultures, the superior holds the power, and respect for hierarchy is strong. Governance systems of projects immersed in cultures with high power distance are mostly unable to constrain abuses. Then, power distance is a predictor of project governance.

There is evidence of cultural differences or subcultures across ranks in projects. While hierarchical position determines rewards, privileges and discrimination are social rank deviations. In terms of privileges, status is a title to social respect. It refers to the honorary attribution of social rank whose outcomes can be depicted as privileges rather than achievement-based honors, conceded to and relished by high-status players in a social structure.

Status derives from social beliefs related to an assortment of social or cultural schemes or stamps (such as class, ethnicity, and family). Merton's "Matthew effect" states that "once a system of social stratification has been established, it is likely to be perpetuated independently of merit" (Washington & Zajac, 2005, p. 285). Cohan (2014) described "the power of the elite" as characterized by factors such as ambition, elitism, prowess, aggressive sexual conduct, racial prejudice, prosecutorial authority, impunity, and perceptions of entitlement and privilege, where people involved in the case remained unwilling, or unable, to comprehend their failures.

French and Raven (1959) described five bases of power: reward, coercive, legitimate, referent, and expert. It can be said, however, that the sources of power in projects and organizations are authority, personal characteristics, expertise, and opportunity. Actions of individuals in organizations, including attitudes and behaviors, are related to their positions and to the attributes of such positions. Power resides in the position, not in the officeholder. Still, in the referent or expert cases, the position limits its realms. Going beyond a vertical (military-type) stratification of positions, it can be considered the position inside a social network. In agreement with this view, Pryke (2005, p. 927) said: "Actions of actors in organizations can best be explained in terms of their position within networks of relationships." He proposed the use of two measures for governance comparative studies: actor point centrality, which provides a measure of the power or prominence of a certain actor; and network density, which provides a measure of connectivity among actors.

The status of an individual within the organization is an element that may help create opportunities for unethical behavior. "The person-situation interactionist model emphasizes the reasoning facet of moral decision-making, acknowledging the role of both individual and situational variables when dealing with an ethical dilemma" (Lopez, 2015, p. 208). "Most managers look to others and to the situation for hints about what is right and what is wrong behavior and how to behave, instead of adhering to their internally-held decisions of right and wrong" (Lopez, 2015, p. 212).

Although cognitive moral development guides the cognitive process of deciding what is right or wrong in a particular situation, situational moderators also influence ethical decision-making behavior. "Among the individual moderators were ego strength, field dependence, and locus of control. The situational moderators include organizational culture (normative structure, referent others, obedience to authority, responsibility for consequences), characteristics of the job (role taking, resolution of moral conflicts), and immediate job context (reinforcement, other pressures)" (Lopez, 2015, p. 208).

When talking about the ethics of responsibility, Drucker (1986, p. 254) claimed: "Men and women do not acquire exemption from ordinary rules of personal behavior because of their work or job. Nor, however, do they cease to be human beings when appointed vice-president, city manager, or college dean." This statement implicitly recognizes the existence of the *Line of Impunity* by denying it based on moral grounds. It clearly highlights the perception that work or job rank may detach people from behaviors expected at lower levels.

Stark (1993, p. 38) asked: "Why should managers be ethical?" Classical political theory and early behavioral studies were not adequate to understand the concept of corporate power, as its locus, ethics, and complexity were not clearly addressed. What is required is a description of the relationship between power and values in organizational life in order for business ethics to provide its normative framework.

When referring to governance, the Sustainable Governance Indicators (SGI, 2015) comprise two dimensions: executive capacity and executive accountability. Under the *Line of Impunity* view, as executive capacity grows, executive accountability fades away, as was manifested in Don Phillips's (Managing Director of Morningstar, Inc.) testimony before the U.S. Senate Committee on Banking, Housing, and Urban Affairs. He remarked:

> . . . *the recent scandals make it abundantly clear that too many people in this industry are willing to forsake their responsibility in exchange for short-term personal profit. Sadly, these were not the acts of a few, low-level employees, but instead were violations of trust that took place at the highest levels, including company founders, CEOs, portfolio managers, and several current or former members of the Investment Company Institute's Board of Governors* (Davis et al., 2007, p. 331).

Referring to the practice of settling fraud cases with companies while not charging employees, Dennis Kelleher, quoted by Schmidt and Wyatt (2012, p. B1) said: "If you are an executive, you know that the chances of getting caught are infinitely small, and the chances of getting caught and prosecuted are even smaller," inferring that the rarity of criminal charges may encourage executives to challenge legal limits. No fear of punishment for ethical violations is then a cultural variable strongly correlated to the *Line of Impunity*.

Seventy-five percent of Americans believe that corruption has increased over the last three years (2012–2015). While trust in big business is going down, the wrongdoing no longer surprises—corporate misconduct has become an everyday event. The perception is that irresponsible corporate behavior exists, and it is widespread.

The surge in penalties assessed by the U.S. Department of Justice reflects a regenerated emphasis on corporate fraud, resurrecting questions about the proportional lack of charges against executives at the companies that are experiencing the starchiest punishments. Senator Jack Reed (D, RI), quoted by Schmidt and Wyatt (2012, p. B1), said: "A lot of people on the street, they're wondering how a company can commit serious violations of securities laws and yet no individuals seem to be involved and no individual responsibility was assessed."

The general public as well as government officials have been calling for more individuals to be held responsible, but still comparatively few prosecutions of individuals have resulted from the biggest fraud cases. Some believe it is much too hard and costly to find evidence linking individual actions to corporate wrongdoing. Dayen (2015) said: "One of the defining issues of this millennium has been the bifurcation of the criminal justice system, with one set of rules for ordinary people and another for elites. We've learned that justice is a commodity to be purchased rather than a universal value delivered without prejudice."

Emulating Friedman's postulate, Porter (2012, p. B1) said: "Company executives are paid to maximize profits, not to behave ethically. . . . Evidence suggests that they behave as corruptly as they can, within whatever constraints are imposed by law and reputation."

During recent decades (since World War II) a diverse group of men and women were elevated into power because of their own degrees and merit. As a result of their success in corporate business, a new privileged class has emerged as new nobility. These people embrace a sense of social distance, basically as an out-of-touch new elite, able to manipulate things in order to stay on top. Brooks (2012, p. A23) said: "The meritocracy has not fulfilled its promise." While the social contract between ordinary citizens and the elite was broken, the expected "merit-based" system is not in fact fair: Actually promoting inequality, it has been dysfunctional, leading to governance failure. "This elite class lacks the awareness of its larger social role, as the value of success substituted the value of virtue. As striving aggressively for success is encouraged by our customs, many people are compelled by determination to succeed at all cost, even by sacrificing ethics" (Lopez, 2015, p. 212).

Decisions in the area of business are based on strategy rather than on ethics. Trevino (1986) proposed that (a) "in actual work related situations, a lower moral judgment can be expected from managers than in response to hypothetical dilemmas" (p. 608), and (b) "external pressures of time, competition, costs, and resources, influence negatively managers' ethical behavior" (p. 614). High executive rewards, and the difficulty of monitoring top management activities, are the proper ground for ethical dilemmas.

Stanford's (2004, p. 14) remarks on the Enron bankruptcy showed that:

> . . . while those at the top of the company were responsible for its collapse, they had sold enough of their stock to bank away tens and in some cases hundreds of millions of dollars. The decent people who worked under them would lose not only their jobs, but in some cases their life savings including the funds in their 401(k) plans that they had invested in Enron stock.

This case constitutes a clear indication that some people were above, and some other people were below, the *Line of Impunity*. Stanford (2004, pp. 15–16) also referred to the existence of power "abusers who often rule with unrestrained autonomy," reinforcing the idea of unleashed conduct above the *Line of Impunity*.

In the quest for ethical compliance, top management can provide legitimate authority; still, Stanford (2004, p. 17) asked: "But is this an ethical catch-22 to ask the ones most responsible for the current malfeasance to now lead the way in supporting compliance to more stringent codes of ethics?"

Top management must be responsible for ethical compliance, because of its high position in the organizational structure and its legitimate authority. In a vertically structured organization, authority is centralized in the top apex of the organization. The higher the centralization, the greater the power and decision-making autonomy. Stanford (2004, p. 20) claimed:

> . . . a mounting grass roots effort led by the Ethics Officers Association calls for every organization to have a chief ethics officer (EO) to set ethical standards and advise upper management. Although these proposals have laudable objectives, they are hard to execute because, in each case, the proposal hinges on commitment and compliance from top management, the same group responsible for the ethical crisis.

It has been argued that ethical decision making and organizational structure are related by the locus of authority and the opportunity to pursue one's own interest. Some people see a positive relationship between the individual's level of authority and the potential for opportunistic action. Conflict theorists support that while protecting their own interests, people with power define deviance to fit their needs. Still, a pluralist organization will develop over time into an oligarchy (a bureaucracy ruled by a few), whose members are motivated to maintain their privileges and power.

The likelihood of financial misrepresentation increases with top management inducements along with poor corporate performance proportional to aspirations, as they provide imperativeness that can encourage firms to act unethically. Some arguable explanations about the diffusion of corruption range from globalization to growing income inequality, to the role of corporations in politics. Political and economic efficiency incentives may have increased the ambit of corporate wrongful conduct as well as the motivation for business executives to bend or violate the laws and precepts.

The organization of corrupt individuals is a phenomenon in which some processes facilitate corrupt behaviors mainly for personal benefit (Pinto et al., 2008). Such an organization can be characterized as corrupt when these corrupt behaviors cross a critical threshold. These concepts parallel at an organizational level what the *Line of Impunity* means at an individual level. They recognize corrupt behaviors as well as the existence of a critical threshold of malice.

Milgram (1975, p. xi) said: "Behavior that is unthinkable in an individual . . . acting on his own may be executed without hesitation when carried out under orders." Schaefer (2005, p. 174) added:

> *Milgram pointed out that in the modern industrial world, we are accustomed to submitting to impersonal authority figures whose status is indicated by a title (professor, lieutenant, doctor) or by a uniform (the technician's coat). Because we view the authority as larger as, and more important than the individual, we shift responsibility for our behavior to the authority figure.*

Even after the compliance systems developed to improve oversight, the regulators' efforts, and the rules enacted after the Madoff affair, problems persist. Theft, Ponzi schemes, and other financial scams continue to happen. Apparently, the Sarbanes-Oxley Act has not had any noticeable effect on the occurrence of corporate unethical behaviors.

Fraud has been verified in 90 percent of the multinational companies or large local companies operating in Latin America, most of it being perpetrated by high-hierarchy personnel. Levine (2005, p. 31) referred to

> *. . . the tendency in some huge corporations (and some less enormous) to believe that they are so successful, so large, so invincible, so much part of modern society, that they are not subject to the same scrutiny by the public that makes any other company answerable.*

In addition, Porter and Russell (2004) found that:

> [Corporations] *operate above many of the laws that ordinary men face and have more rights than American citizens; also that they are complex entities that act according to their own autonomous set of motives and dynamics, and they are capable of doing great harm in the pursuing of self-interest.*

In the same line of thought, Nace (2003) perceived two sets of rights: one for corporations and another for citizens. He also perceived that laws that apply to citizens do not reach corporations, which is the essence of the *Line of Impunity*.

The *Line of Impunity* was described almost a century ago by F. Scott Fitzgerald in his book, *The Great Gatsby,* when a hit-and-run episode involving people from different social levels was ignored by high society as if basic rules of care and decency did not apply to them.

> *. . . I saw that what he had done was, to him, entirely justified . . . —they smashed up things and creatures and then retreated back into their money or their vast carelessness or whatever it was that kept them together, and let other people clean up the mess they had made . . .* (Fitzgerald, 1925/2013).

The *Line of Impunity* was also demonstrated by Tiger Woods's (Woods, 2010) comments after his extramarital scandal: (a) "I knew my actions were wrong, but I convinced myself the normal rules didn't apply. . . . I felt I was entitled"; and (b) "Money and fame made me believe I was entitled."

5.6 Focus Group Session #2: The Line of Impunity (Lopez, 2015, pp. 227–231)

Free Association

The first question was

> *What words or phrases come to mind when you think about "The Line of Impunity"?*

Responses

> *Integrity, line not to be trespassed upon.*
> *Politicians avoid this line, or they could be in trouble.*
> *Beyond could be illegal or unmoral.*
> *Government making laws for some to be above and others below.*

The second question was

> *Line of Impunity refers to the idea that certain ranks or positions in the social hierarchy are entitled to prerogatives or advantages, and that the power granted at those levels transcends the limits of control or law enforcement (such as a tyranny)* [Lopez, 2015, p. 205]. *Could you comment?*

Responses

> *The Line of Impunity generates a sense of power to bend the rules and increases the latitude an individual has in his job.*
>
> *In the military ranks comes with privileges, but you need to have integrity and responsibility.*
>
> *You know there is a line not to cross.*
>
> *In the corporate world if you know an executive level person, you should better please his administrative assistant as she has the power to make your life great or miserable.*

The third question was

> *According to Washington & Zajac, (2005, p. 285), "Hierarchical position determines rewards." Could you comment?*

Responses

> *In the military it does. You have right to get a house according to your position. Officers have better houses. In corporations it is a little flatter except when you talk about money.*
>
> *Higher positions have better rewards, but they come with higher risks too.*
>
> *In the corporate world, the higher up, the more perks but also the higher the risk of losing them.*

The fourth question was

> *Stanford (2004, p. 19) said "A positive relationship exists between the individual's level of authority and the potential for opportunistic action to increase one's personal interests; that is, the higher the level, the higher the potential return when opportunistic actions are taken." Could you comment?*

Responses

> *Executives could use their knowledge for opportunistic advantages.*

Executives can use inside information.
Government should exert its power to stop this kind of abuse.
People such as Martha Stewart do take advantage of opportunities.

The fifth question was

> *"Conflict theorists point out that people with power protect their own interests and define deviance to suit their own needs. Income tax evasion, stock manipulation, consumer fraud, bribery and extraction of kickbacks, embezzlement, and misrepresentation in advertising—these are all examples of white-collar crime, illegal acts committed in the course of business activities, often by affluent, 'respectable' people"* (Schaefer, 2005, p. 186). *Could you comment?*

Responses

> *Because there is no law against it, it is not illegal to take advantage.*
> *People think they can get away with their mess, look at Anthony Weiner in New York.*
> *There was an executive in Alabama who went to jail due to mortgage fraud. He argued that his CEO views couldn't be confronted, so he accommodated his own.*
> *There is a percentage of people that truly believe that they are above others and that they are doing nothing wrong, and there is another percentage of people that know they are wrong but believe they are entitled. They believe they can get away with it.*
> *There is peer pressure to protect the interests of the group.*

Role-Play

Statement 6: Managers at lower levels in the organizational structure experienced more pressure to accommodate their personal values to attain company goals (Carroll, 1978).

Responses

> *My manager pressured all our team to improve the score of customer service satisfaction surveys asking us to explain to the customers that we need to get an A grade and instructing us to pressure them in this direction instead of collecting bias-free feedback. It was against my personal beliefs but I couldn't confront him.*
> *I can sense that if we were under pressure to meet the quota, an accommodation of values could have happened.*

Statement 7: "As higher education bears a relationship to levels in the organizational hierarchy, moral discernment will increase with hierarchical level" (Trevino, 1986, p. 609).

Responses

I disagree. Perhaps it was true 20 years ago, but not now.
You learn morals from your parents; education is not a factor in moral judgment.
Education provides skills but you cannot learn morals in school.
At higher levels you navigate politics, so you have to bend morals.
The environment could entice you to bend the rules.

Final Comments

Rank comes with privileges.
Privileges and risks have a positive correlation.
Opportunistic actions are possible at higher ranks as privileged information is available.
Peer pressure is used to maintain privileges.
An accommodation of values can occur below the Line of Impunity.
It is not clear whether education can improve moral discernment.

5.7 Summary of the Line of Impunity

The focus group discussion, along with the survey and the research on the Line of Impunity can be summarized as follows.

People associate the *Line of Impunity* with the qualities of integrity and responsibility; it is recognized as a line not to be trespassed on, as it could be illegal or unmoral.

While there is skepticism about politicians and government integrity (*government making laws for some to be above and others below*), there is also a claim for government involvement (*government should exert its power to stop this kind of abuse*).

The main driver of the *Line of Impunity* seems to be associated with the balance between privileges and rewards on one side of the scale and risks on the other side. A positive correlation between privileges and risks is recognized. As higher positions are related with higher risks, their privileges are accepted (*rank comes with privileges*).

Beyond the *Line of Impunity*, a sense of power to bend the rules is perceived, increasing the latitude an individual has in his job and fostering the exploitation of opportunistic advantages and the use of privileged information (*people such as Martha Stewart do take advantage of opportunities*). There is a feeling that above a certain position in the hierarchy, different rules apply (*at higher levels you navigate politics, so you have to bend morals*) and that the context is a factor (*the environment could entice you to bend the rules*).

There is a percentage of people that truly believe that they are above others and that they are doing nothing wrong, and there is another percentage of people that know they are wrong but believe they are entitled. The first group is victim of ethical fading due to elitism, in that they do not realize their posture is inappropriate (unconscious transgression), while the latter group chooses to follow the unethical path (conscious transgression). This group of conscious transgressors believes they will always come away clean (*people think they can get away with their mess*).

It is also noticed that the reaction of those individuals who cannot confront someone above the *Line of Impunity* is accommodation (*. . . he argued that his CEO views couldn't be confronted, so he accommodated his own. It was against my personal beliefs, but I couldn't confront him. I can sense that if we were under pressure to meet the quota, an accommodation of values could have happened. An accommodation of values can occur below the Line of Impunity*).

The value of education related to the *Line of Impunity* is not fully embraced (*it is not clear whether education can improve moral discernment*), as some people believe that these kind of morals are supposed to be nurtured at home (*you learn morals from your parents; education is not a factor in moral judgment*).

Chapter 6

Ethical Issues and Ethical Dilemmas

The first responsibility of project managers is generally regarded as project delivery in response to their project sponsors' demands, with the main objective of getting the job done. This view risks reducing project managers' role to simply following orders. Rather, because project managers' decisions involve people and resources, and might impact the environment, they require ethical commitment, going beyond the mere delivery of the project. Conflicts that might affect project cost, schedule, risks, safety, and quality of deliverables may generate ethical issues throughout the project life cycle.

As discussed in Chapter 2 with reference to "double standards," there are at least two sets of personal moral values, like two contradicting brains. In the case of individuals working on a project, they should have one project brain and another, non-project brain. Ludlow (2013, p. 2) said: "The mid-level managers that he spoke with were not 'evil' people in their everyday lives, but in the context of their jobs, they had a separate moral code altogether." Ludlow's statement confirms that some decisions are based on project considerations (job ethics), while others are based on a dissimilar private ethical code.

If we accept that people may have various sets of ethical values (not just two) and that these sets can change or evolve over time, it is difficult to picture personal virtue as the sole indicator of project managers' ethical probity. Personal virtue is not enough. Hints of projects' ethics from past experiences and psychological tests should be considered in order to obtain a more accurate indication of project managers' job ethics.

As an example, Figure 6-1 shows how this works for an individual with three sets of ethical values—one for home, another for the project, and a third one for a vacation in Las Vegas with college friends. While the horizontal axis depicts the ethical values (lust, gluttony, greed, sloth, wrath, envy, and pride), the vertical axis indicates the level of moral intensity required to be enticed to violate each ethical value. It can be observed that the levels of moral intensity required to violate the home ethical value set are higher than those required to violate the project and the vacation ethical value sets.

In a utopian world, no enticement of any kind should modify one's stand on ethical matters. Nevertheless, this utopian view does not consider the degree to which passion can influence people. When fascinated by passion, emotions may confuse the boundary between right and wrong. "Even the most brilliant and rational person, in the heat of passion, seems to be absolutely and completely divorced from the person he thought he was" (Ariely, 2008, p. 99).

"The ethical decision-making process is an issue-contingent matter, meaning that the characteristics of the moral issue (moral intensity) are important determinants in this process" (Lopez, 2015, p. 41). While Dante's seven deadly sins (lust, gluttony, greed, sloth, wrath, envy, and pride) can be considered an early basic framework for ethical issues, some lines of inquiry in the study of projects and society relations include project ethics, project social responsibility, ideology, attitudes, values, project social performance, stakeholder engagement, minorities, gender, community, international affairs, consumers, employees, environment, and interaction with the government.

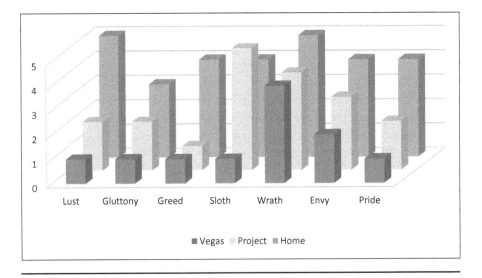

Figure 6-1   Various Sets of Ethical Values

Renz (2007, p. 132) put together a table of "integrity challenges," identifying paternalism, disregard for stakeholders, conflict of interest, public pressure, strategic opponents among partners and subcontractors, informal structures, collusion, structure-inherent loyalty issues, bypass agreed responsibilities, incompetence, corruption, indiscretions, defamation, fake ingratiation, mobbing and bullying, hidden agendas, and team abuse.

Some project conduct generally considered improper include environmental violations, production of dangerous consumer goods, misleading advertising, deceptive sales practices, violations of government labor regulations, white-collar crime, theft, fraud, embezzlement, vandalism, sabotage, absenteeism, withdrawal, withholding effort, sexual harassment, unethical decision making, and injustice. As society is not static, social issues evolve over time, together with changes in the perception of these concepts. People today are concerned about social issues such as consumerism, environmentalism, discrimination, product safety, occupational safety, and stakeholder issues.

Project ethical issues may be related to the following areas (Lopez, 2015, pp. 60–71):

1. Values: conflicts of interest and personal honesty
2. Rights: treatment of women and minorities, employee rights, sexual harassment, equal employment opportunity, and whistle-blowing
3. Operations: financial procedures, accepted business practices in foreign countries, workplace safety, and environmental issues
4. Deliverables: product safety and quality
5. Society: stakeholder interests, project contributions, social issues, public disclosure, and customer privacy
6. Government: government officials and government contracts

Although a rigorous alignment and matching of the above-mentioned conduct is difficult and could be subject to further discussion, a summary of the different views on ethical issues might include: consumerism, product safety, occupational safety, discrimination, and environment issues (Carroll, 1978); sexual harassment, theft, fraud, unethical decision making, embezzlement, absenteeism, withdrawal, vandalism, and sabotage (Harper, 1990); interpersonal abuse, conflict of interests, personal decadence, security of company records, financial procedures, bribery, appropriation of others' ideas, taking unfair advantage, lying, buying influence, violating rules, product quality, workplace safety, giving or allowing false impressions, abuse, privacy, and advertising content (Jennings, 1999); corporate social responsibility, business ethics, minorities and women, ideology, attitudes, values, and stakeholder management (Whetten et al., 2007); deceptive sales practices, violation of labor regulations, and production of dangerous consumer goods (Harris & Bromiley, 2007); "taking office materials and supplies, submitting false project status and progress reports, submitting false budget information, accepting offers for kickbacks or gifts from suppliers, overcharging of project work hours, falsifying expense reports, knowingly implementing poor quality products, and approving inaccurate test results" (Durham, 2010).

Furthermore, the magnitude of the ethical affair does not relate to its rightness. Fraud can range from large cases, such as embezzlement of funds and assets or fraudulent financial reporting, to minor cases such as employee theft and unproductive behavior (*Small Sins Allowed*).

In addition, individuals are generally risk-averse when facing gains and risk-disposed when facing losses. Poor performance is associated with a project's pursuit of higher risks. Under pressure, any gimmicks which anticipate achievement of a goal are permitted. Project managers may try to achieve their aspirations via legitimate means when they are close to their reference points, while they may find few lawful alternatives when performing far below their aspirations. Thus, when a project's performance is better than aspirations, the probability of facing an ethical issue is generally lower. On the other hand, projects not meeting aspirations are the proper scenario for a desperate response such as misrepresentation.

Kish-Gephart et al. (2010) mentioned "bad apples" (individual characteristics), "bad cases" (moral issue characteristics), and "bad barrels" (organizational environment characteristics) as the factors affecting the unethical choices (see Table 6-1). They found that honest people may act unethically under specific

Table 6-1 Unethical Choices (Intention/Behavior)

Individual Characteristics		Moral Issue Characteristics	Organizational Environment Characteristics
Psychological	Demographic		
Cognitive moral development	Gender	Concentration of effect	Egoistic ethical climate
Idealism	Age	Magnitude of consequences	Benevolent ethical climate
Relativism	Education level	Probability of effect	Principled ethical climate
Machiavellianism		Proximity	Ethical culture
Locus of control		Social consensus	Code of conduct
Job satisfaction		Temporal immediacy	Code enforcement
		General moral intensity	

Adapted from Kish-Gephart, J. J., Harrison, D. A., and Trevino, L. K. (2010), Bad Apples, Bad Cases, and Bad Barrels: Meta-Analytic Evidence About Sources of Unethical Decisions at Work. *Journal of Applied Psychology*, Vol. 95, No. 1, p. 3.

circumstances, and that factors beyond the person's dispositions to act can affect behavior.

Another view was presented by Renz (2007, pp. 45–46) that identified an ethical issue "in the form of a gap in an integrative and complete institutionalization of the ethical discourse." This is the case when relevant ethics policy topics are not properly discussed at the project level; when ethical issues faced by the project team are not discussed, analyzed, and meditated upon; when the experiences with ethical issues at the project level are not shared upward; and when the ethical considerations of policy development are not explained downward to the implementation levels.

6.1 Moral Neutralization

A basic assumption is that "people do not ordinarily engage in reprehensible conduct until they have justified to themselves the rightness of their actions" (Bandura et al., 1996, p. 365). In connection to this, Sykes and Matza (1957) identified five techniques of moral neutralization: denial of responsibility, denial of victim, denial of injury, condemnation of condemners, and appeal to a higher loyalty.

When denying responsibility, a claim is often made to forces beyond control. Confronting these forces, the individual feels powerless. As a result, people transfer the guilt of their actions to another entity. In projects it is common to find reports of delays due to the weather, workers' strikes, supplies shortages, and other unpredictable reasons, although in many cases the real scenario shows a lack of proactivity in the handling of these factors. *Small Sins Allowed* also plays a role in denying responsibility when the moral intensity of the issue is lessened such that responsibility for such a small affair is not worth being considered (such as walking over an ant). "Because we are so adept to rationalizing our petty dishonesty, it's often hard to get a clear picture" (Ariely, p. 219).

By denying injury, people minimize that any damage or hurt could be done, frequently appealing to a higher good, where the consequences of their acts are minor and can be disregarded. This is often seen in large and complex projects, as people working on such projects may believe that small acts of dishonesty, including withholding information and misrepresenting facts, cannot hurt the overall project because of its magnitude or complexity. Nevertheless, as noted by Tom Peters as referenced by Renz (2007, p.113), "there is no such thing as a minor lapse of integrity," which aligns with *Small Sins Allowed*.

Denial of victim occurs when people acknowledge that their actions might have some negative impact, but they think those affected do not merit protection because they have only themselves to blame. "When employees felt exploited by the company . . . these workers were more involved in acts against the organizations as a mechanism to correct perceptions of inequity or injustice" (Hollinger & Clark, 1983, p. 142).

Condemnation of condemners takes place when people accuse their judges of lacking understanding of the praxis in which they are engaged. By raising doubts about the motivations or ideology behind such criticism, moral worries are bent back toward the critics. "People who are critical of a decision to be less than truthful in a project may be silenced by claims to the effect that they have no idea of what it is like to work in the heated atmosphere of this project" (Kvalnes, 2014, p.597). This is how the *Line of Impunity* frequently works.

When appealing to a higher loyalty, people may refuse to be guided by self-motivation, invoking a prominent duty conceived as more significant than any moral value. These allegiances might be related to family, friends, colleagues, employer, country, religion, etc. In any project, higher loyalties are detrimental, as decisions can be made, and actions can be taken, not by obeying reason and morals but because of allegiances. A necessary condition for project success identified by Turner and Müller (2004) is that all the project participants should view the project as a partnership, within which their objectives are aligned, which is managed to achieve the best results for all.

Projects should nurture a culture where attempts at neutralization are not consented to. In this regard, project managers should work to generate an internal communication climate that allows project team members to speak up when facing morally challenging actions, as silence is an accomplice of neutralization.

6.2 Ethical Hierarchy

The major ethical issues in projects fall into three categories.

The first ethical issue concerns the negative impact of project deliverables. Is the project delivering something that affects people or communities negatively? In such a case, is this something that could be justified in view of a larger positive effect or a higher good (e.g., although generating moderate noise and pollution, the project provides employment to the local community)?

The second ethical issue refers to the relation between the project and its parent organization and its related stakeholder community. Often this ethical issue is related to transparency and information sharing. What are the consequences of being completely transparent? Could the project be damaged or even closed? On the other hand, what are the harms of not disclosing enough information in a timely fashion?

The third ethical issue is internal within the project. Here we find areas such as whether or not to follow. Sometimes it may appear to be more effective not to follow the strict letter of all the processes, cutting some corners, or diverting resources from the area they were assigned to another area without proper approvals. Should it be acceptable to take into consideration different points of views and/or scenarios by applying different rules? Is the project manager's goal the project's goal, or does he or she have his or her own agenda?

The way of acting in relation to those ethical issues depends on various aspects such as the individual moral and values, the processes and policies, and the level of ethical knowledge and awareness.

Similar to Maslow's (1943) needs hierarchy theory, we defined a hierarchy for ethical issues according to their moral intensity. Although Maslow's theory establishes invariable levels in a pyramid, in our case we do not believe that our ethical hierarchy is cast in stone, as many factors might alter it. Still, it is a valid starting point to understand the source of individual responses to these issues (see Figure 6-2).

The reaction to ethical issues of low moral intensity (work deviance or unprofessional conduct) comes from individual morals. This constitutes the basic level in the ethics hierarchy. Here is where the *Small Sins Allowed* reside. People use their moral baggage (what they bring from learning at home, school, church, or society) to resolve ethical issues of low moral intensity. The stronger the individual moral formation, the better decisions can be expected when facing ethical issues of low moral intensity.

The response to ethical issues of medium-low moral intensity (favoritism or disrespect) comes from policies, processes, and procedures. This is the second level in the ethics hierarchy. People find that their moral baggage is not enough to make a sound decision because of the complexities of the situation, so they find support and guidance in the policies, processes, and procedures in place; ". . . intuitive ethical behavior is not enough for the integrity of a project" (Renz,

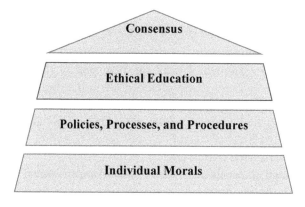

Figure 6-2   Ethical Hierarchy

2007, p.118). More solid, easy-to-follow, and consistent project policies, processes, and procedures can lead to better decisions when facing ethical issues of medium-low moral intensity.

The answer to ethical issues of medium-high moral intensity (harassment or discrimination) comes from ethical education. This is the third level in the ethics hierarchy. People use what they have learned from specific ethical training, or from their experience of previous similar situations, to resolve ethical issues of medium-high moral intensity. Facing the same ethical issues of medium-high moral intensity, more effective ethical decisions can be expected from individuals with stronger ethical education than from others lacking it. In this regard, no one can expect that a single isolated ethical training class can generate an adequate level of ethical awareness. Ethical education must be recursive, updated, and pertinent.

The reply to ethical issues of high moral intensity (corruption) comes from consultation, deliberation, and consensus. This is the highest level in the ethical hierarchy. At this level, neither the individual morals, the project policies, processes, and procedures, nor ethics education are enough to support an ethical decision. The moral intensity of the ethical issue is so high that one individual alone cannot handle it properly, making it necessary to collegiate the decision making by: consulting with those who have the experience, wisdom, or authority; deliberating with peers and team members; and gaining consensus among stakeholders.

Knowing these possible reactions, responses, answers, and replies should help us select adequate project manager and project team members; assemble effective project policies, processes, and procedures; select a sound project ethics training program; and establish a project governance process that ensures easy access to consultation, deliberation, and consensus (see Case 6-1).

Case 6-1 Complimentary Breakfast

A new expense policy was introduced in a large corporation with headquarters in Sweden. Under the new policy, the daily meals allowance for business travel included all meals (breakfast, lunch, and dinner). Consequently, in order to avoid double expensing, all employees had to deduct from their daily meals allowance the amount of any meal charged as part of the hotel bill or expensed as business meals with customers.

Although all hotels in Sweden offered complimentary breakfast, Hans (the project manager) requested his project team members to follow the new policy by subtracting a certain amount from their daily meals allowances. Other project managers in the same program did not request that their project team members deduct any such amounts. As a consequence, Hans's project team members complained, as they felt unhappy and discriminated against. Why were the deductions applied only to them and not to other projects' team members? Hans thus faced two problems. First, his project was affected negatively, as project team members started to refuse to work with him, asking for transfers to other projects. Second, his professional career was harmed, as his manager gave him feedback saying that he lacked flexibility.

Hans was in the presence of an ethical dilemma. Following the new policy was unpopular and negatively impacted his project and his career. Not following the new policy should result in double expensing, which was exactly what upper management was trying to avoid. How could this happen?

The answer to questions related to ethical dilemmas are always different, depending on the lens we use to look at the question. One extreme posture considers that all the rules must be followed, all the time, to their full extent. The opposite posture considers that complimentary breakfast should not be counted as a daily meal, as "complimentary" does not have a value attached to it (although there were guidelines saying that breakfast represent 20 percent of the daily allowance). This was a kind of norm in almost every Swedish company—something that "everybody did." A middle-level posture would ask: "Is bad always wrong?"

Questions

1. As breakfast was complimentary, was there a valid point in assuming a monetary value to be deducted from the daily meals allowance? Is that fair?
2. What if all meals were complimentary; should employees still be allowed to claim the daily meal allowance?
3. Was this a case of *Small Sins Allowed*?
4. Do you believe that it is better to have some flexibility?
5. Did Hans cross the *Line of Impunity*, acting at his own discretion regardless of other project managers' interpretations?
6. What happens when customary usage, including *Small Sins Allowed*, changes due to new rules?
7. Could this case have been resolved by individual morals alone, or by policies, processes, and procedures?

> 8. Could ethics education have helped to resolve this issue?
> 9. Was this a case where deliberation, consultation, and consensus were required?
> 10. Could you provide examples from your own experience?

If we believe that there are good and bad people, it is important to ward off the incorporation into the different project ranks of those with dubious ethical backgrounds. Nevertheless, if we believe that most people confronted with a conflict of interest can cheat, then conflicts of interest must be eliminated, as avoiding temptation is easier than surmounting it. To be reassured about the project manager's capability to handle ethical issues within the project, both aspects must be addressed by staffing the project team with sound ethical personnel and by avoiding any possible conflict of interest.

Project policies, processes, and procedures should be revised to provide a coherent, updated, and easy-to-follow set of tools. Extensive and overlapping policies add to confusion. Consistency of project policies, processes, and procedures is necessary in order to guarantee their effectiveness.

It has been observed that although some people cheat, given the chance to think about honesty (moral reminder), they stop cheating. When people are given no ethical benchmarks, they tend to fall into dishonesty, but if they are regularly reminded about morality and job ethics, then they are much more likely to be honest (see Case 6-2). For this reason, specific ethics training must be provided to the project manager and the project team on a regular basis in order to reinforce their predisposition toward ethical behaviors. Nevertheless, we must bear in mind that ethics training alone is an old-fashioned way of education that does not always generate ethical awareness. Other, more imaginative approaches should be tried, such as performing activities that act as moral reminders, generating slogans, and using personalized signs and posters with funny situations and pictures of actual employees. One approach is to select monthly an "Act as One,", showing examples of real people who acted in an exemplary way and awarding them with some token of appreciation to encouraging others to follow their example.

Case 6-2 Moral Reminder

Kvalnes (2014, p. 595), mentioned:

> *Mazar et al. (2008) set up an experiment to test the honesty of students. They wanted to explore whether cheating among respondents could be affected by moral reminders. A total of 229 students*

participated. They were asked to perform math tasks and were given opportunities to cheat when reporting on the results of their individual performances. Before the test, the respondents were asked to write down either the names of ten books they had read in high school (no moral reminder) or the Ten Commandments (moral reminder).

. . . if character is the most influential factor on moral behavior, only insignificant differences should be observed between the two groups. The outcome, however, was that the respondents in the first group of students showed normal cheating behavior, while all the respondents in the second group refrained from cheating all together. Evoking the Ten Commandments served as a moral reminder, and cheating was eliminated.

Ariely (2010, p. 288) has conducted a similar experiment where respondents were asked to sign a statement to the effect that they understood that what they were about to do fell under a university honor system. The result was the same as with the Ten Commandments. The act of signing served as a moral reminder, and made them refrain from cheating.

Questions

1. Could you tell why moral reminders as honor codes and codes of ethics actually reduce dishonesty in projects?
2. It was observed that moral reminders immediately before the test worked; nevertheless, when the moral reminders are static (a code of ethics exists but it is just a piece of paper), they are forgotten and their effect may fade. Should the project code of ethics be revisited periodically? How often?
3. Elaborate about your own project experiences with and without moral reminders.

Project managers require easy access to an ethical collegiate body for consultation, deliberation, and consensus. This body need not necessarily be part of the project, as small projects cannot afford to have one. Instead, the parent organization can sponsor this ethical collegiate body to provide support to many projects. An ethical compliance officer can help in establishing project policies, processes, and procedures, in implementing ethics training programs, and in serving as a member of the collegiate body.

Table 6-2 Ethical Issues

Project Deliverables	Relation between Project and Organization/Community	Internal to the Project	Magnitude	Response
Ignore/hide safety threats (1)	Pollution (4)	Corruption (8)	High moral intensity	Consultation, deliberation, consensus
Lying about compliance with emission standards (2)	Unsustainability (5)			
Hiding defects and bad product quality (3)	Fraud (6)			
	Bribery (7)			
		Discrimination	Medium-high moral intensity	Ethics education
		Harassment		
		Favoritism	Medium-low moral intensity	Policies, processes, procedures
		Disrespect		
		Unprofessionalism	Low moral intensity	Individual morals
		Work deviance		

(1) General Motors ignition-switch lawsuit.
(2) Volkswagen diesel emissions cheating scandal.
(3) Toyota lied about deadly acceleration glitches.
(4) Exxon Valdez oil spill.
(5) Aggressive overfishing.
(6) Enron.
(7) Siemens.
(8) Fédération Internationale de Football Association (FIFA).

By devoting careful attention to the circumstances that surround ethical issues, whether they are internal to the project, related to the relation between the project and its parent organization or its community, or relative to the project's deliverables, and by providing adequate means to foster healthy responses at any moral intensity level, there is a greater likelihood of attaining honest conduct.

Table 6-2 shows the correlation among ethical issues, the moral intensity associated with them, and the source of their responses. As the same ethical issue can have different moral intensities depending on many factors (e.g., while a small bribery may be considered as having low moral intensity that can be addressed by individual morals, harassment and discrimination may be considered as having high moral intensity). For simplicity, we took only one value for each ethical issue/moral intensity pair. In consequence, Table 6-2 should be considered only an example of the many possible combinations.

6.3 Ethical Dilemmas

Ethical dilemmas arise when project managers face the need to make decisions misaligned with their own values (the project considerations collides with their personal ethical belief system). In their reasoning process, project managers are concerned about the project survival, the consequences of their decisions, their reputation, and their responsibility. In the resolution of ethical dilemmas, a reliance on past ethical experiences, instincts, backgrounds, and judgment are major factors.

Müller et al. (2013, p. 36) investigated four types of ethical dilemmas.

Dilemma type 1: "There is a conflict between two equally valid ethical choices." This dilemma is commonly resolved by choosing the better good or the lesser bad; still, the choice may leave a bitter taste. This is a daily challenge in today's projects. As many factors interact, it is not uncommon that some or many of them have demanding requests that collide with others. Choices may have many facets, and the best choice may still have flaws.

Dilemma type 2: "There is a conflict between what is ethically correct and what the company policy is." This is a common type of dilemma in certain industries, which is due to outdated policies or to the nature of company business in the face of changing society values over time, such as in the tobacco, coal, and nuclear energy industries.

Dilemma type 3: "There is a conflict between what is ethically correct and what the law dictates." This type of dilemma is commonly seen in international projects, where cultural differences confront what is legally right in one country with what is ethically unacceptable in another.

Dilemma type 4: "There is a conflict between what is legally correct and company policy." In this case the policy is against the law (e.g., "creative accounting"). There may also be some gray areas where conflicting laws (consider different countries) allow companies to choose the more convenient policy.

Adding some findings from their research, Müller et al. (2014, p. 45) described seven categories of ethical dilemmas affecting projects: "transparency [concerned with the correctness of progress reports that may be affected by uncertainty, fear, and hope], relationship, optimization, power and politics, illegal actions, role conflicts, and under-performing governance structure." Each of these ethical dilemmas deserves due consideration, as the way project managers address them can make the difference between project success and project failure.

6.4 Summary

Integrity challenges arise because people may have various sets of ethical values. Under specific circumstances, even honest people may act unethically. Nevertheless, people need to feel their actions are correct or at least justified, which is done by denying responsibility, denying of victim, denying injury, condemning the condemners, and appealing to a higher loyalty (moral neutralization).

People's behavior depends on individual moral values, the processes and policies, ethical knowledge and awareness, and even consultation and consensus. It can be better understood with the help of the "ethical hierarchy" discussed in this chapter, which accounts for successive levels that require the use of different means according to their moral intensity.

Still, ethical dilemmas arise when project managers face the need to make decisions that may be misaligned with their own values.

Chapter 7

The Ethics Cube

Ethical theory does not offer a decision procedure as ethics cannot be reduced to a system of rules (Bredillet, 2014, p.551).

7.1 Basic Relations

The interaction between ethical behaviors and project governance is reciprocal and influenced by the context that surrounds it. This relationship can be depicted as a double-headed arrow pointing to the ethical behaviors at one head and to the project governance at the other head, floating inside a context bubble (see Figure 7-1).

Although many people think that governance influences ethical behaviors, not otherwise, this vision is skewed and partial. It is true that without a

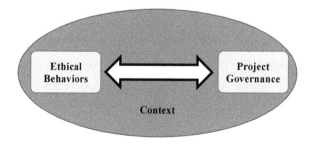

Figure 7-1   Ethical Behaviors and Project Governance

governance system one cannot expect to have a frame for ethical behaviors, but we believe that any governance system comes to life based on the ethical behaviors and expectations of its constituents. It is also true that governance systems evolve according to the evolution of the ethical behaviors of the society in which they are immersed.

The desired or expected ethical behaviors that fit into the particularities of the project context are used as building blocks to build the project governance system, which in turn helps to guard those ethical expectations against misbehavior.

Now that we know about the *Small Sins Allowed* and the *Line of Impunity*, it is time to explore their relationship in order to better understand where they fit into the big picture and their relationship with project governance. *Small Sins Allowed* and the *Line of Impunity* are only two of the many potential ethical behaviors, so when focusing only on this pair of behaviors we are looking from the general into the particular. In this regard it is important to understand that these are not the only possible behaviors, so we should not disregard the importance of other ethical behaviors.

Three facts were observed in our research. First, when the *Small Sins Allowed* increase in magnitude or number, the quality of project governance decreases, and vice versa. Second, the relationship between *Small Sins Allowed* and project governance is affected by the *Line of Impunity*—meaning that keeping all other factors the same, when the *Line of Impunity* has been trespassed, the impact of *Small Sins Allowed* in project governance increases. Third, the context intervenes in the relationship among *Small Sins Allowed,* the *Line of Impunity,* and project governance. In return, the quality of project governance restrains the occurrence of *Small Sins Allowed,* the trespassing of the *Line of Impunity,* and influences the project's context. This model in which each of the participants is connected to the others by double-headed arrows is depicted in Figure 7-2.

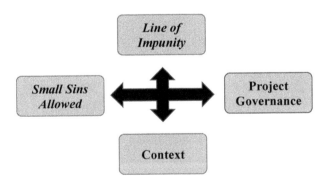

Figure 7-2   Model

If we ask how the *Small Sins Allowed* affect project governance, the answer is that the recurrence of *Small Sins Allowed* (work deviance, disregard for the rules, acceptance of mild abuse, etc.) acts to the detriment of the project's ethical standard as the confrontation attitude of project managers is worn out, the objectives definition becomes blurred, and the monitoring and control capability fades away. In a project where work deviance, mild abuse, and disregard for the rules are systematically tolerated, the steering capacity dissipates, ultimately affecting the chances of project success, which become erratic.

In longer projects, the acceptance or customarity of *Small Sins Allowed* has a spiral effect that keeps "small," growing over time, in such a way that something not tolerated at the earliest moments of the projects becomes tolerated eventually (see Case 7-1). On the other hand, consistent project governance initiatives can render long-term benefits to the project by elevating the thresholds of *Small Sins Allowed* and the *Line of Impunity* (see Case 7-2).

Case 7-1 Sleeping Beauty

Once upon a time, the manager of a project overslept. Although he was a competent professional and very technically skilled, he consistently arrived late for work. During the initiation phase he was never more than 30 minutes late, and he usually stayed late when required, so the project team felt condescending toward him and tried to show up for work at least a few minutes before him, making them consistently 15 minutes late.

By the planning phase, The manager's delays grew to 45 minutes to 1 hour. Although he blamed the traffic during the morning commute, his team nicknamed him "Sleeping Beauty." Team members started to follow the trend, and no one showed up before 8:30 a.m.

When execution started, the normal was between 1 hour and 90 minutes late, which made 9:00 a.m. the regular start of the day for the project team. It was during this phase that problems arose with contractors who were paid by the hour and expected to start at 8:00 a.m., and with third parties that charged waiting time for the cranes and trucks that were ready to work early in the morning when no one from the project team had yet shown up.

In order to cope with this situation, flex time was introduced as a palliative. It came into the project not because someone evaluated its benefits but because there was no other way to cover the project manager's *Small Sin* of tardiness. It added complexity to the project schedule, as same tasks couldn't start without all the team

---

members and contractors allocated to them. Delays and additional cost required re-baselining.

Eventually, a small weakness in the project manager's character ended up as a big project concern that eroded his authority, cost extra resources, and delayed the project.

Questions

1. Do you believe that it is valid to allow a certain flexibility, say, 15 minutes?
2. Is there a way to cope with the growth of *Small Sins Allowed* like this? Explain.
3. How do you depict a project where more than one *Small Sins* are allowed?
4. Can you provide examples from your experience?

Case 7-2 A City Turnaround

Once upon a time there was a city with high crime rates (New York in the 1970s), where the police were focused on fighting major crimes such as assault, rape, and murder, while ignoring less serious offenses such as fare beating, tagging (graffiti), and panhandling. Life was miserable for regular citizens, who felt unsafe and surrounded by unruly individuals and gangs in control of the streets.

One day the Mayor and the Police Commissioner decided to launch a sanitation project. They started to clean up the streets by focusing on those crimes where there was not an individual victim but that affected the whole neighborhood, causing fear and misery to the residents.

It was realized that the serious crimes were to a certain extent intangible to most people, as they did not confront murder in their everyday lives. What they faced every day and what concerned them was related to visual and noise pollution, order in the streets and subways, and a general perception of insecurity.

Before this sanitation project, thieves, panhandlers, prostitutes, pimps, and all kinds of antisocial individuals felt over the *Line of Impunity*, as the police didn't have eyes on them. Eventually, attention to the minor offenses paid off, as the community started to feel safer and happier having a governance system in place that focused on their daily life struggles, such as noise, prostitution, and the general appearance of the streets. There was also a perception of living in a better city where rules were obeyed and where the law was in control of the situation. Significantly, the serious crime rate fell more than 50 percent.

Questions

1. Do you believe that nothing is insignificant and that solving the *Small Sins Allowed* can positively impact the outcome of a project? Explain.
2. Do you believe that lowering the *Line of Impunity* can positively impact the outcome of a project? Explain.
3. What is your position regarding "zero tolerance" policies? Explain.
4. Can you provide examples from your own experience?

The effect of the *Line of Impunity* in the relationship between *Small Sins Allowed* and project governance can be explained by the change in mindset that the perception of the *Line of Impunity* generates in all the project ranks, which facilitates the accommodation to lax standards. When project managers and project team members feel entitled to prerogatives, they will likely be enticed into unruly conduct. Although this conduct might include plain fraud and deception, the most common and mildest enticement relates to raising the threshold of the *Small Sins Allowed*.

There is not a single or universal way to measure the impact of *Small Sins Allowed* and the *Line of Impunity* on project governance, as the context intervenes in this relation. Anomie, moral intensity, cultural differences, pressure and opportunity, work environment (stress, communications, permissiveness), and scarceness–abundance are context-related factors that affect the weight of the *Small Sins Allowed* and the *Line of Impunity* on project governance (see Case 7-3).

Case 7-3 Alfa Romeo Alfasud

In the 1960s, the Alfa Romeo was a technologically advanced and beautifully designed masterpiece of a car. In 1971 Alfasud came to life—a brilliantly designed car fitted with Alfa Romeo's first boxer engine, with a low center of gravity and front disc brakes. *Autocar* magazine voted it the best front-wheel-drive car ever built.

In spite of its initial prospects, there were clouds in the horizon. Alfa needed to compete in a growing segment dominated by Fiat, so the Alfasud was built as an inexpensive car aimed at the masses by an intrusive Italian government (owner of Alfa Romeo at the time), who resolved to open a factory in the largely agricultural and poor south of the country, in order to help its industrialization.

When production started, it suffered from quality problems: door gaps weren't even, interior bits weren't screwed together very well, and the body was made of the notoriously inferior Russian steel. The combination of an unskilled and unmotivated workforce, cost-cutting initiatives, and inferior steel quality resulted in new cars rolling off the production plant with rust on them.

As a result, the Alfasud was doomed by its poor reputation, which some attributed to the fact that it was not built in a factory in Milan, "where it should have been built."

Questions

1. Could you find *Small Sins Allowed* in the failure to attain quality standards due to an unmotivated workforce? Where else?
2. Could you find the *Line of Impunity* in the government intrusion that led to the decision to cut cost by using low-quality materials? Where else?
3. Explain the context differences between the traditional factory in Milan and the new one in Naples.
4. Which project governance initiatives could have helped to restrain the negative influence of the *Small Sins Allowed*, the *Line of Impunity*, and the context? Explain.

7.2 The Elements of the Ethics Cube

7.2.1 Ethical Values

Ethics relates to customs, habits, and practices, guiding moral decisions on what actions are right or wrong in specific contexts, situations, or domains. "People may have various sets of ethical values" (Lopez, 2015, p. 51), such as professional ethics, family ethics, and general ethics (see Figure 6-1). These sets of ethical values can change because of situational or temporal factors. Each of these sets endorses some ethical issues by recognizing the level of moral intensity required to overcome them.

Professional ethics are related to: conflicts of interest, honesty, women and minorities, employee rights, sexual harassment, equal employment opportunity, financial procedures, accepted business practices in foreign countries, safety, environmental issues, quality, stakeholder interests, project contributions, social issues, public disclosure, customer privacy, government officials and government contracts, among others.

Family ethics includes honesty, loyalty, industry, faith, loving, supporting each other, protecting children from obscenity and exploitation, and a position about abortion, abstinence education, premarital sex, polygamy, and recreational drugs, among others.

General ethics includes wisdom, prudence, fortitude, temperance, justice, goodwill, balance, and a position about liberty, free speech, pursuit of happiness, democracy, slavery, war, human rights, environment, social inclusion, and racism, among others.

7.2.2 Personal Interests

In this setting, personal interests are related to our own goals and tastes, and with the intensity with which we pursue them. It distinguishes two factors: the feeling that engages our attention, concern, involvement, or curiosity with something, and the power of our will to reach those goals and to satisfy those tastes.

Some interests are trivial, and some others are complex. They span from simple wishes that we expect to get without effort, such as an aromatic coffee in the morning before starting our day, to fiercely pursued hard desired objects, such as our ambition to become the project manager of a large international endeavor.

Our interests recognize two origins or sources: one is biological and the second is induced. Biological interests are mainly related to our tastes, as comfortable shoes, spicy food, cold drinks, soft skin, nice smell, sunny days, starry nights, and love. Induced interests are majorly related to our goals and ambitions, such as travel, sports, money, power, and success. There is plenty of room to discuss whether some interests are biological or induced, as many of them have elements of both.

The importance of induced interests is well recognized by marketing professionals who struggle to impose stereotypes as they recognize that most people do not know their wishes unless they see them in context. As we always look at the things around us in relation to other things, we revolve around the comparative advantage of one thing over another.

Sometimes, personal interests have the power to bend our ethical standards, modifying some or all our sets of ethical values.

7.2.3 Allegiances

By allegiances we recognize the loyalty or devotion to some person, group, cause, nation, religion, or the like.

The concepts of "herding" (assuming other people's behavior) and "differential association" (how the exposure to postures favorable to vicious acts leads to the infringement of rules) were introduced in Chapter 2. It was also noted that although acceptable conduct can be ruled, people with influence over us, such as friends, fellow employees, comrades, priests, leaders, and the like, may promote rather dissimilar rules. According to Merton (1938), certain individuals engage in nonconformist conduct because of the pressure exerted by social structures.

Allegiances are always culturally induced and once embraced are hard to change. They determine our behavior and in some cases take precedence over personal interests in the sense that we are able to sacrifice them as allegiances belong to a higher level of devotedness. Likewise, if there is a conflict between ethics and allegiances, it is not uncommon that the latter take precedence.

7.2.4 Opportunity

> *Managing to see reality from a self-serving perspective is not an exclusive moral flaw, limited only to "bad people." It's a common human foible and is part of being human* (Ariely, 2008, p. 293).

Recalling the times of the founding fathers of the United States, we could all agree that they were individuals of the highest moral character; nevertheless, like most people of their time, they fell prey of a self-serving view of reality. Some embraced slavery as normal, depriving people of other races of common human rights, even citizenship. A century ago, during the construction of the Panama Canal, Americans were paid in gold, while others were paid in silver. In present days and across borders, the treatment of domestic help is still skewed toward the self-serving perspective of the employers, who in many cases pay less than minimum wages and do not recognize their employees' rights to social security, vacations, and other benefits.

We can depict two types of dishonesty. One evokes the image of thieves, deceivers, and cheaters; another is the kind perpetrated by individuals who ordinarily consider themselves to be honest people.

Ariely (2008, p. 202) said that "individuals are honest only to the extent that suits them." Given the opportunity, honest people tend to cheat. Nevertheless, they may cheat just a little bit. It seems like an internal honesty monitor turns on only when they think about large transgressions. Aligned with *Small Sins Allowed*, they do not even consider how these little transgressions affect their honesty.

There are three conditions for dishonesty to take place: motivation (positive in the form of incentives or negative in the form of pressure), opportunity, and rationalization that there is nothing wrong in what is being done. However, sometimes motivation and rationalization are not observed, and dishonesty still occurs. This fact makes some people believe that opportunity is the only relevant factor for dishonesty. The absence of controls or the ability to override them provides the opportunity. The agency theory supports that individuals will act opportunistically (illegally) if there is inadequate monitoring.

7.3 The Ethics Cube

We can imagine a cube such as the Rubik's Cube,* with one face for each of the elements discussed above: professional ethics, family ethics, general ethics, personal interests, allegiances, and opportunity. Each face consists of a number of squares, with each one representing an attribute of this element.

The faces will be named according to their elements as follows:

Professional: professional ethics
Familiar: family ethics
Citizen: general ethics
Selfish: personal interests
Idealist: allegiances
Pragmatic: opportunity

In its original state, each face shows exclusively the attributes of its own element.

A brief list of attributes includes:

Professional: effectivity, efficiency, quality, timely delivery, cost consciousness, team building, leadership, risk management, and integrity

---

* Rubik's Cube is a 3-D combination puzzle invented by Ernő Rubik, originally called the Magic Cube. It shows one different color in each face; the game consists of bringing them back to the original state after shuffling. It is considered to be the world's best-selling toy.

Familiar: loyalty, supportiveness, faithfulness, caregiving, tenderness, and love

Citizen: good manners, politeness, fairness, justice, inclusiveness, trustworthiness, social involvement, environmentalism, and tolerance

Selfish: wrath, greed, sloth, pride, lust, envy, and gluttony

Idealist: membership, submission, accomplice, and the common good of the party members

Opportunity: free ride, easy gains, wild card, shortcut, fame, and power

7.3.1 The Shuffled Ethics Cube

We said that the ethics cube is shuffled when not all its six faces are in the original ideal state where each square of the faces represents an attribute of the corresponding element (see Figure 7-3). Instead, elements of different faces are mixed up, providing a distribution of attributes that lacks uniformity.

The shuffling happens when conflicts arise from our attempts to balance incompatible goals. Then, if we look at the Professional, Familiar, and Citizen faces, we will find some of the attribute squares replaced by squares coming from other faces such as the Selfish, Idealist, or Pragmatic.

> *On one hand, we want to look in the mirror and feel good about ourselves.
> . . . On the other hand, we're selfish, and we want to benefit from cheating.
> . . our flexible psychology allows us to act on both of them when we cheat
> "just by a bit"* (Ariely, 2008, p. 292).

Small Sins Allowed is the mechanism that allows us to cheat on our own values "just by a bit"; otherwise, replacing too many Professional faces' squares

Figure 7-3 Ethics Cube

by Selfish, Idealist, or Pragmatic ones might compromise our own professional self-esteem. Nevertheless, as discussed before, the *Line of Impunity* has an effect, making people believe they are entitled to perquisites, with the effect of an increase in the magnitude and/or volume of the *Small Sins Allowed* without affecting professional self-esteem, as the feelings of entitlement back up these actions.

The context plays an intervening role, making it easier or harder for the *Small Sins Allowed* to take place and for the *Line of Impunity* to moderate them—as if, according to the context, the movements of our cube become either more lubricated or stiffer.

7.3.2 The Professional Face

The Professional face gathers together all the attributes of professional ethics. When referring to a particular project, it shows the attributes of project ethics.

In a shuffled situation, this face loses some of its attributes as conflicting attributes from other faces take precedence over them. By not seeing evil in dishonesties of small moral intensity, or by tolerating them, what we are doing is replacing some attributes of our ideally defined project ethics by attributes belonging to our own interests, our allegiances, or just by simple opportunity attributes. For every attribute missing (every square in our Ethics Cube), the project performance and its chances to succeed are affected negatively. In addition, when the *Line of Impunity* is trespassed, more squares will be replaced from the original Professional face by those coming from other faces, accentuating the denaturalization from the ideal professional ethics.

Then, the aim of every project governance initiative should be to restrain the cube from changing its original ideal state, by watching for the small transgressions, avoiding the feeling of impunity, and creating an adequate context for the project. No project success is possible when project ethics are jeopardized by personal interests, allegiances, or opportunities to cheat.

7.3.3 Selfish Shuffling

When a selfish attribute takes over a professional attribute, we are said to be in the presence of a selfish shuffling. It is not uncommon to find examples of greed taking over risk management, envy taking over team building, lust taking over leadership, or gluttony taking over quality.

One illustrative example is Richard Grasso's massive and unreasonable compensation (over $130 million) during his tenure between 2000 and 2002 as

CEO of the New York Stock Exchange. Although his case was dismissed in court, it is an everlasting public memory of how personal greed can take over an institution's prestige and reputation (Thomas, 2006).

Another example is provided by Michigan State Representatives Todd Courser and Cindy Gamrat (both members of the state legislature's Ethics Committee and both married to other people), who lied about their extramarital affair after being caught "making out" inside a car in a park. While the first resigned, the latter was expelled from the state legislature. This, of course, was not the only case in which the lust of public servants overtook their leadership (Lawler, 2015).

7.3.4 Idealist Shuffling

Idealist attributes can shuffle with professional ones—for example, membership with efficiency, submission with leadership, or the common good of party members with cost consciousness. In these cases, allegiances take priority over the project's success.

In a family business it is not uncommon that a member of the family, such as a son or nephew, jumps into the company ranks via the sole merit of his relationship to the owner. From the perspective of the owner, nothing other than his or her own economic interest is affected. Nevertheless, from the perspective of other employees of the firm, who had job growth expectations and better qualifications than the appointed relative, it is viewed as unfair and frustrating.

The situation is even worse in public entities, where the economic interest of the owners is managed by agents whose allegiances may be seriously detrimental when they favor people who share their political or religious views over other, more qualified candidates.

7.3.5 Opportunity Shuffling

Opportunity attributes can shuffle with professional ones, such as easy gains with risk management, power with leadership, fame with team building, shortcuts with quality, or wild cards with integrity.

We can find plenty of examples of project managers who take all the credit for a project milestone attained and thereby alienate their project team members, or project managers who prepare their project's schedules by reducing the number of trials, sacrificing quality of the final deliverable and adding risks to the project.

7.3.6 Other Shuffling

There are also cases where ethical values are attributes of other faces, such as Family or Citizen, which can shuffle with professional ones. In these cases, even when these attributes are ethical, they do not belong to the professional realm, denaturalizing its essence, as, for example, when the good citizen attribute of tolerance is wrongly applied in a project, relaxing the project's standards.

7.4 Summary

As people's professional ethics are affected by factors that belong to other realms such as personal and general ethics, as well as interests, allegiances, and opportunity, it is necessary to understand that professional ethics is not a theoretical field in which only project-related elements should be considered.

It is also of capital importance to understand how the *Small Sins Allowed* and the *Line of Impunity* facilitate the shuffling of factors among the faces of the ethical cube. Having them restrained improves project governance, limiting the shuffling and making every move difficult.

The aim of any governance system should be to solve the ethical cube (its six faces of a solid color each). The better the project governance, the sooner the ethical cube is solved.

Chapter 8

Final Words

Perhaps, beguiled by custom and order, one's sense of evil went numb (Kawabata, 1969, p. 19).

8.1 Findings and Definitions

In this final chapter, we will briefly revisit some findings and definitions presented in the previous chapters, such as social and cultural structure, ethical virtues and principles, moral judgment, workplace deviance, *Small Sins Allowed*, double standards, legality, context, collective patterns of thinking, perception of values, project culture, *Line of Impunity*, governance, project governance, and a discussion about the benefits of rules and principles.

In an orderly world, everyone gets what he or she deserves; good is rewarded, and bad is punished. According to Merton (1938), two elements of the social and cultural structure are worth considering: the culturally defined goals and interests, and the regulations and controls imposed when describing the acceptable modes of achieving these goals. Social groups, such as corporations and projects, match their goals with the regulations that provide permissible ways of achieving them.

Even though ethical virtues concern the superior "end" of improving the lives, well-being, and happiness of society at large, as managers overcome the *Line of Impunity,* a detachment between business and ethics results in confusion, as many believe themselves above the moral law. Their own inner logic causes a disconnect between economic reasons and moral purpose, departing from the ethical values of the society. Nevertheless, "society, organizations, and projects should not necessarily pursue different goals, as what is good for one does not need to be bad for the other and vice versa" (Lopez, 2015, p. 31). The integrity of common purpose is an integral component of project management, as achieving project objectives requires an alignment with the prevailing public belief. Some people have observed that this is true in larger and complex projects, although for small projects they found it hard to visualize this alignment because of the limited scope of these projects.

"Fundamental ethical principles are integrity (honesty, sincerity, and candor), justice (impartiality, sound reason, correctness, conscientiousness, and good faith), competence (capable, reliable, and duly qualified), and utility (quality of being useful and to provide the greatest good for the greatest number)" (Lopez, 2015, p. 35). The cornerstone of project ethics must include honesty, good faith, transparency, responsibility, respect, and fairness. Project ethical issues are related to values (conflicts of interest and personal honesty), rights (treatment of women and minorities, employees' rights, sexual harassment, equal employment opportunity, and whistle-blowing), operations (financial procedures, accepted business practices in foreign countries, workplace safety, and environmental issues), deliverables (product safety and quality), society (stakeholder interests, project contributions, social issues, public disclosure, and customer privacy), and government (government officials and government contracts).

Recognizing what is ethical does not mean doing it, as moral judgment alone is not enough for moral action. Decisions are based on emotions, ego strength, courage, personal interests, allegiances, and opportunity. Other factors such as competition, pressure, different cultures, and managerial values can lead to unethical choices. In this way, dishonesty may be the result of neutralizing initial moral dissonance as well as of finding excuses to act against moral convictions. Some kinds of "creativity," such as risk taking, rule breaking, conflict creation, challenging authority, and sacrificing long-term value for short-term rewards, are ethically objectionable behaviors.

"Workplace deviance (violation of organizational norms) differs from ethics (right or wrong behaviors in terms of justice, law, and morality)" (Lopez, 2015, p. 89). Minor workplace deviances are mainly *Small Sins Allowed* (leaving early, taking excessive breaks, working slowly, wasting resources, showing favoritism, gossiping about co-workers, or blaming co-workers), while ethical misbehaviors include sabotage, accepting kickbacks, lying, stealing, sexual harassment, verbal abuse, or endangering co-workers.

"There are those who associate deviant behaviors with inexperience, and those who associate deviant conducts with the power granted at higher levels of the hierarchy. There is a perception that deviant conducts are more common at senior levels than at junior levels" (Lopez, 2015, pp. 197–198), indicating the negative influence of the *Line of Impunity*. While trying to infer the perceptions of organizational business ethics by hierarchical levels, Ardichvili et al. (2012) found three levels: Executives on top perceive their organization ethical business culture as very positive, managers have a less positive view, and employees have the lowest positive perception. It was also found that time within the organization increased the positive perception of executives, while it decreased the perception of managers and employees. They explained that executive perceptions were related to protecting the organization and their own status quo, while managers and employees were critical and psychologically distanced.

There is an indication that an issue falls into the category of *Small Sins Allowed* when: (a) the moral issue is not recognized leading to an unconscious transgression; or (b) the moral issue is recognized, but it is deemed too small for serious consideration, leading to a conscious transgression. For one reason or another, *Small Sins Allowed* are excluded from the moral intent and from the moral behavior, "representing a buffer zone between accepted and reprobated behaviors. Rather they lie between permissiveness and conformity, where permissiveness does not necessarily mean socially-accepted behavior, but socially-tolerated behavior instead" (Lopez, 2015, p. 200).

Subjective morality is one of the main drivers of the *Small Sins Allowed*, associated with multiple ethical mindsets, moral intensity, cultural differences, and managers' willingness to justify ethically suspect behaviors. Other drivers of *Small Sins Allowed* are uncertainty avoidance, the meaning of time, and tolerance for deviant ideas and behaviors. Also, the exposure to postures favorable to vicious acts (differential association) positively predisposes the commission of *Small Sins Allowed*, particularly "everybody else does it" and "never mind the rules."

Legality does not always involve the totality of an issue's perceived morality. Objective morality corresponds to promulgated laws, while subjective morality corresponds to beliefs regarding the appropriateness or incorrectness of an action (moral sense, scruples, and conscience). Double standards (a deviation between words and reality) affect projects negatively, as employees appreciate honesty in communications. Donaldson (1996, p. 5) pointed out that "Even though most large U.S. companies have both statements of values and codes of conduct, many might be better off if they didn't. Too many companies don't do anything with the documents; they simply paste them on the wall to impress employees, customers, suppliers, and the public. As a result, the senior managers who drafted the statements lose credibility by proclaiming values and not living up to them." When a forbidden pattern flourishes in a project without consequences, it becomes a "de facto" expectation that changes the meaning of acceptable behavior.

Context determines what values are accepted, and its influence in projects is evidenced in the cultural, political, and business environment in which the projects operate. As ethical decision making is context-specific, morals can be influenced by situational variables. Then, project governance practices must evolve with the circumstances, complexity, and changes, adapting to the specific context.

"Collective patterns of thinking [such] as the meaning attached to life, values, beliefs, and the distinction between good and evil, true and false, beautiful and ugly are components of a culture" (Lopez, 2015, p. 99). Culture establishes which behaviors are acceptable and which are not. It should not be associated exclusively with geographies (countries or regions), since there are other cultural spaces such as company, industry, profession, function, and project.

The perception of values is affected by four superimposed drivers: global, local, industry, and project. Values vary from culture to culture, so there is no culture-free project and managing a project is culturally dependent. Then, project culture is the set of expectations of conduct that act on the members of the project. "Integrity, ethics, and responsibility are demanded in the contemporary workplace, based on a growing acknowledgement that good ethics is a socially shared value" (Lopez, 2015, p. 79). When studying the intercultural aspects of business, Donaldson and Dunfee (1999, p. 45) said: "Global managers often must navigate the perplexing gray zone that arises when two cultures—and two sets of ethics—meet." They maintained that as no one can claim to own a universal set of ethical norms, tolerance toward others' ethics should be promoted. Their "Integrative Social Contracts Theory" (pp. 52–53) indicates five sets of norms:

1. *Hypernorms (fundamental human rights or basic prescriptions common to most major religions)*
2. *Consistent Norms (culturally specific* [but still] *consistent both with Hypernorms and other legitimate norms, including those of other cultures)*
3. *Moral Free Space (norms that are inconsistent with at least some other legitimate norms existing in other cultures)*
4. *Group Norms that often express unique, but strongly held, cultural beliefs*
5. *Illegitimate Norms (norms that are incompatible with Hypernorms)* [containing those] *values or practices* [that] *transgress permissible limits*

According to Ardichvili et al. (2009, p. 446), ethical business culture is "based on alignment between formal structures, processes, and policies, consistent ethical behavior of top leadership, and informal recognition of heroes, stories, rituals, and language that inspire organizational members to behave in a manner consistent with high ethical standards that have been set by executive leadership." Along the same line of thought, Donaldson (1996, p. 6) said: "A company's leaders need to refer often to their organization's credo and code and must themselves be credible, committed, and consistent. If senior managers act as though ethics don't matter, the rest of the company's employees won't think they do, either."

Ardichvili et al. (2012, p. 423) considered as aspects of an ethical business culture, trust, integrity, honesty, leadership effectiveness, stakeholder balance, mission, vision, values, and process integrity. They also named some "formal elements of an ethical culture, which encompass: 1) the commitment of the organization to establish and sustain a compliance program through the adaption of ethics standards and procedures; 2) taking appropriate measures to establish a system for reporting misconduct; and 3) providing the appropriate safeguards to ensure that employees use the reporting system."

Some aspects of Ross and Benson's (1995, p. 350) culture variables applied to project culture are potential sources for ethical misconduct, such as a "reward structure [careless of ethical limits], [a rationalization that] 'everyone plays the game', ambiguous ethical rules and [deficient] training, inability to comprehend that ethics are a legitimate constraint [of projects], and no [concern about sanctions] for ethical violations." No fear of punishment for ethical violations is strongly related to the *Line of Impunity*. Some people see a positive relationship between the individual's level of authority and the potential for opportunistic action. They believe that people with power define deviance to fit their needs, such as maintaining their privileges and power.

The main driver of the *Line of Impunity* is associated with privileges/rewards and risks. A positive correlation between privileges and risks is recognized, as higher positions are related to higher risks. Beyond the *Line of Impunity,* "a sense of power to bend the rules is perceived, increasing the latitude an individual has in his or her job" (Lopez, 2015, p. 228) and fostering the exploitation of opportunistic advantages and the use of privileged information. There is a feeling that different rules apply above a given level in the hierarchy ("at higher levels you navigate politics, so you have to bend morals") and that the context is a factor ("the environment could entice you to bend the rules"). "There is a percentage of people that truly believe that they are above others, that they are doing nothing wrong (unconscious transgression), and there is another percentage of people that know they are wrong but believe they are entitled (conscious transgression)" (Lopez, 2015, p. 229).

Governance possesses three main characteristics: legality (lawfulness), legitimacy (power is linked to a mandate from the parties involved), and participation (having a role and sharing responsibilities). Other desirable traits are moral purpose, accountability, sustainability, transparency, responsiveness, consensus orientation, equity, and inclusiveness. According to Johnstone et al. (2006), its components are "meanings" as authority structure (attaches decision-making responsibility to positions), and "means" as mechanisms (processes and methodologies required to implement decisions) and policy (dissemination of information).

Project governance aims for transparent, repeatable, and scalable principles, structure, and processes. Intrinsically, it should fit the project's context, setting the framework for accountabilities, strategic alignment, decisions, roles, management, control, and ethics. As project governance can change because of external forces (economy, legislation, technology, social pressure, or religion) and internal forces (policies, processes, or leadership), it requires steady review to maintain its effectiveness and quality.

Effective project governance is a fundamental requirement for project success. The nature of project governance needs to be related to business ethics (legitimacy), processes and procedures (legality), and behavior and practices (participation). "Some commonly agreed-upon principles are: regard for the stakeholders, integrity and ethical conduct, transparency and opportunity in the disclosure of information, and executive control" (Lopez, 2015, p. 117).

Formal conventions become internalized by education and enforcement. Then, while soft external pressure can stimulate the internalization of virtues, strong pressure might have the opposite effect, generating forced compliance, distrust, and revolt. A demand to codify common principles and expectations is associated with the complexity of the project. Rules that classify acceptable and unacceptable behavior are typically easier to follow than principles, which may allow considerable latitude. Nevertheless, rules can be complex, outdated, and circumvented.

In project governance, some advocate for the application of rules, while others advocate for the application of principles. Rules may be easier to dictate and comply, making a clear distinction between acceptable and unacceptable behaviors, but they reduce the project manager's discernment. Rules are positive for static situations and bureaucratic top-down structures. They serve as references to deal with the status quo; nevertheless, changes such as technological breakthroughs and new processes, represent a serious challenge for rules. Also, even well-enforced rules can still be circumvented. In contrast, principles allow for tailored decision for each situation under the umbrella of the wider scope of project integrity. Principles may avoid flaws of outmoded rules. Donaldson (1996, p. 6) stated: "Striking the appropriate balance between providing clear direction and leaving room for individual judgment makes crafting corporate values statements and ethics codes one of the hardest tasks that executives confront."

Projects tend to be governed at all levels by applying the same school (rules or principles), which may be counterproductive. Rules may be a good fit for the project team while restraining the project manager's ability to cope with uncertainties. Principles may be good for the project manager while allowing the project team to find their own way instead of following standards. Instead, a mixed approach (where some positions in the project are best suited to rules, while others are best suited for principles) should be considered. Positions concerned with the project work seem to be a good match for rules, while higher-level positions concerned with the achievement of project outcomes seem to be a better fit for principles.

A top-down approach, where ethics belongs to corporate governance and spills down to projects, may work for long lasting projects with strong bonds to the organization; nevertheless, for contemporary project work that involves a large number of heterogeneous contractors and third parties, corporate ethics spillage is difficult. A project culture is not always built based on corporate spillage, as the different cultures of its members are hard to mix during the relatively short project's time span. In such a context, the project manager's ethical standpoint is the force that glues the shared values of the stakeholders.

8.2 How to Improve Governance in Projects

People have various sets of ethical values, and those sets evolve over time as changes in the perception of social issues do. For this reason, personal virtue as the sole indicator of project managers' ethical probity is not enough.

Kish-Gephart et al. (2010) mentioned bad apples (individual characteristics), bad cases (moral issue characteristics), and bad barrels (organizational environment characteristics) as the factors affecting the unethical choices. Under pressure, projects managers performing below their aspirations tend to take higher risks, permitting any actions which anticipate success (apples). Even honest people may act unethically under specified circumstances, and factors beyond the person's dispositions to act can affect his or her behavior. Projects not meeting expectations are the proper scenario for deception (cases). For this reason, the project management offices (barrels) should pay special attention to those underperforming projects. These projects should be the primary target for training, audits, moral reminders, and consultation.

Projects' major ethical issues are related to (1) the negative impact of project deliverables, (2) the relationship between the project and both its parent organization and its related stakeholder community, and (3) internal within the project. The way these ethical issues are faced depends on the project manager's values, sound processes and policies, and the level of ethical knowledge and awareness of the project team.

As in Maslow's (1943) needs hierarchy theory, there is a hierarchy for ethical issues according to their moral intensity, which suggests responses to these issues (see Figure 6-2). Issues of low moral intensity (work deviance, unprofessional conduct) constitute the basic level in the ethics hierarchy. Here is where the *Small Sins Allowed* reside. People use their moral background to solve these ethical issues. Issues of medium-low moral intensity (favoritism or disrespect) constitute the second level in the ethics hierarchy, since ". . . intuitive ethical behavior is not enough for the integrity of a project" (Renz, 2007, p. 118). People find support and guidance in the policies, processes, and procedures in place. Issues of medium-high moral intensity (harassment or discrimination) constitute the third level in the ethics hierarchy. People use what they learned from ethical training or from their experience as witnesses of previous similar situations to resolve ethical issues of medium-high moral intensity. Ethical education needs to be recursive, updated, and pertinent. Issues of high moral intensity (corruption) constitute the highest level in the ethical hierarchy. At this level the individual's morals, the project policies, processes, and procedures, or ethical education are not enough to support ethical decisions. One individual alone cannot handle these issues properly, making it necessary to use a collegial decision-making process of consultation, deliberation, and consensus.

To be reassured about the project manager's capability to handle ethical issues within the project, two aspects must be addressed: staffing the project team with sound ethical personnel and avoiding any possible conflict of interest. Project policies, processes, and procedures must provide a coherent, updated, and easy-to-follow set of tools. Abundant and overlapping policies add to confusion.

Although some people cheat, if they are regularly reminded about morality and job ethics, then they are much more likely to be honest (moral reminder). For this reason, specific ethical training must be provided to the project manager and the project team on a regular basis. Other imaginative approaches should also be tried, including playing moral reminders, slogans, personalized signs with funny situations and posters with pictures of actual employees who acted with exemplary ethical standards, and rewarding these people with tokens of appreciation to encourage others to follow their examples.

Project managers require easy access to an ethical collegiate body for consultation, deliberation, and consensus. An ethical compliance officer may help in establishing the project policies, processes, and procedures, in implementing the ethical training programs, and serving as a member of the collegiate body.

Honest conduct can be stimulated by committing attention to the circumstances that surround the ethical issues (whether they are relative to the project's deliverables, related to the relationship between the project and its parent organization or its community, or internal to the project) and by providing the means to respond adequately at any moral intensity level.

When the *Small Sins Allowed* increase in magnitude or number, the quality of project governance decreases, and vice versa. The recurrence of *Small*

Sins Allowed (work deviance, disregard for the rules, acceptance of mild abuse, etc.) acts to the detriment of the project's ethical standards as the confrontation attitude of project managers is exhausted, the objectives definition becomes blurred, and the monitoring and control capability fades away. In a project where work deviance, mild abuse, and disregard for the rules are systematically tolerated, the steering capacity dissipates, ultimately affecting the chances of project success.

"The relationship between *Small Sins Allowed* and project governance is [affected] by the *Line of Impunity*" (Lopez, 2015, p. 309), meaning that keeping all other factors alike, when the *Line of Impunity* has been trespassed, the impact of *Small Sins Allowed* in project governance increases. This can be explained by the change in mindset that the perception of the *Line of Impunity* generates in all the project ranks, which facilitates the accommodation to lax standards. When project managers and project team members feel entitled to prerogatives, they will likely be enticed into improper conduct. Although this conduct may include fraud and deception, the most common and mildest enticement relates to raising the threshold of *Small Sins Allowed* (see Case 8-1).

Case 8-1 Internal Allegiance and External Contractor

Back in 2000, during a huge nationwide optimization effort at the main cellular telecommunication provider in Mexico, a workforce of international contractors was hired to fix technical problems such as call accessibility and retainability.

The company found itself incapable of doing that work with their own resources for many different reasons, including lack of experience, insufficient manpower, and its need to focus principally on daily operational issues. What the company failed to recognize was that a culture of apathy and a network of internal allegiances negatively affected all kinds of optimization and change initiatives.

One day, an external contractor requested that the local technical manager send his crew to fix a problem at a particular cell site, by adjusting the position and tilt of the antennas. Some days later the technical manager reported that the work was completed. As the problem continued to be observed, it was requested to confirm that the work was done according to the specifications. The confirmation arrived soon afterwards.

Even after the second confirmation, the statistics associated with that cell site continued to show the same number of connection failures and dropped calls, moving the contractor to check himself. He visited the site, a seven-story building on which the rooftop antennas showed no sign of any work having been done recently. The contractor also noticed some anomalies in the equipment room, which indicated that nobody had visited that site for months.

Once the incident was reported and escalated to the next level of authority, it was found that the local technical management crew had consistently lied about their whereabouts and the work performed. Nevertheless, this conduct was not unknown, and it was tolerated because allegiance to old companions and friendship between the technical manager and his crew was invoked. As a result, the technical manager was only lightly reprimanded, and so his relationship with the proactive contractor was fatally damaged.

By acting proactively and looking for the truth and the quality of his work, the external contractor became a menace for the local team, who were entrenched in their culture, uses, politics, egos, and allegiances. The contractor was not aware of allegiances that made some people feel they were above the *Line of Impunity*, and that opened the door for *Small Sins Allowed*. At the end of the contractor's term, his contract was not renewed.

Questions

1. Do you believe that prerogatives between old companions and friends can be allowed in the workplace?
2. Did the crew feel they were above the *Line of Impunity* because of their relationship with the technical manager?
3. Why do some people feel enticed into improper conduct that raises their threshold of *Small Sins Allowed*?
4. Can you provide examples from your experience?

The interaction between ethical behaviors in projects and project governance is reciprocal. The context intervenes in the relation between *Small Sins Allowed*, the *Line of Impunity*, and project governance. In return, the quality of project governance restrains the occurrence of *Small Sins Allowed*, the trespassing of the *Line of Impunity*, and influences the context.

For each individual the Ethics Cube represents in its six faces the three sets of ethical values (professional, family, and general), the interests, the allegiances, and the opportunity.

"People may have various sets of ethical values" (Lopez, 2015, p. 51), such as professional ethics, family ethics, and general ethics (see Figure 6-1). Each of these sets endorses some ethical issues by recognizing the level of moral intensity required to overcome them. They can change based on situational or temporal factors.

Professional ethics are related to conflicts of interest, honesty, treatment of women and minorities, employee rights, sexual harassment, equal employment opportunity, financial procedures, accepted business practices in foreign countries, safety, environmental issues, quality, stakeholder interests, project contributions, social issues, public disclosure, customer privacy, government officials and government contracts, among others.

Family ethics include honesty, loyalty, industry, faith, loving, supporting each other, protection of children from obscenity and exploitation, and a position about abortion, abstinence education, premarital sex, polygamy, and recreational drugs, among others.

General ethics include wisdom, prudence, fortitude, temperance, justice, goodwill, balance, and a position about liberty, free speech, pursuit of happiness, democracy, slavery, war, human rights, environment, social inclusion, and racism, among others.

Personal interests are related to one's own goals and tastes and to the intensity with which they are pursued. They span from simple wishes to fiercely pursued hard desired objects. Interests can be biological or induced. Biological interests are mainly related to our tastes, such as comfortable shoes, spicy food, cold drinks, soft skin, nice smell, sunny days, starry nights, and love. Induced interests are majorly related to our goals and ambitions, such as travel, sports, money, power, and success. Interests have two characteristics, the feeling (passive) that engages our attention, concern, involvement, or curiosity with something, and the power (active) of our will to reach those goals and to satisfy those tastes. Personal interests have the power to bend our ethical standards.

Allegiances are the loyalties or devotion to some person, group, cause, nation, religion, or the like. According to Merton (1938), certain individuals engage in nonconformist conduct because of pressure exerted by social structures. People with influence over us, such as friends, fellow employees, comrades, priests, leaders, and the like, may promote these behaviors. Allegiances are culturally induced and, once embraced, are hard to change. They determine our behavior, and in some cases take precedence over personal interests in the sense that we are able to sacrifice them because our allegiances belong to a higher level of devotedness. Likewise, if there is a conflict between ethics and allegiances, it is not uncommon that the latter take precedence.

Given the opportunity, even honest people tend to cheat. Ariely (2008, p. 202) said that "individuals are honest only to the extent that suit them." For the little transgressions (*Small Sins Allowed*), people do not even consider how these actions affect their honesty. Some believe that "opportunity is the only relevant factor for [dishonesty]" (Lopez, 2015, p. 71). The absence of controls or the ability to override them provide the opportunity to do so. The agency theory supports that individuals will act opportunistically (illegally) if there is inadequate monitoring.

Exercise 8-1 Drafting Your Own Ethics Cube

For this exercise, your first task is to name eight components for each face of your Ethics Cube.

Professional Ethics

1.
2.
3.
4.
5.
6.
7.
8.

Family Ethics

1.
2.
3.
4.
5.
6.
7.
8.

General Ethics

1.
2.
3.
4.
5.
6.
7.
8.

Personal Interests

1.
2.
3.
4.
5.
6.
7.
8.

Allegiances

1.
2.
3.
4.
5.
6.
7.
8.

Opportunities

1.
2.
3.
4.
5.
6.
7.
8.

Your second task is to build the professional ethics face of someone you know—perhaps your last project manager, a co-worker, or an acquaintance. You can use any component of other faces to which he or she attaches the highest priority or moral intensity, or that he or she cannot resign. In doing so, for each component that you take from other faces, you must drop one of the original components of the professional face named above in order to keep only eight components.

Professional Ethics of John/Jane Doe

1.
2.

3.
4.
5.
6.
7.
8.

Now, as a final task, you need to build your own professional ethics face using the same procedure indicated before.

My Own Professional Ethics

1.
2.
3.
4.
5.
6.
7.
8.

Questions

1. Do John/Jane Doe's professional ethics depart from the pure professional ethics? Why?
2. Do your own professional ethics depart from the pure professional ethics? Explain.
3. In your view, what can be done to restrain interests, allegiances, and opportunities from eroding professional ethics? Explain.

8.3 Guidelines for Ethical Leadership

"Creating a company culture that rewards ethical behavior is essential" (Donaldson, 1996, p. 8). Adherence to the law as the single ethical consideration is not enough. As ethical principles should not compete with having a competitive posture, it is necessary to have a moral imperative through ethical content embedded into corporate vision.

The perception of fairness affects loyalty, commitment, job satisfaction, and trust in management. The lack of motivation or the presence of frustration in the project workplace may drive the violation of norms (workplace deviance).

Project managers can prevent ethical lapses by maintaining their ethical standards aligned with those of the society. Project managers have a strong influence in what project team perceive as *Small Sins Allowed*. For this reason, it is important for project managers to behave ethically and to clearly define what is or is not tolerated. Examples, training, and continuous reminders can help setting the bar.

Projects should nurture a culture in which attempts at neutralization (denial of responsibility, denial of victim, denial of injury, condemnation of condemners, and appeal to a higher loyalty) are not consented to. Project managers should generate an internal communication climate that allows project team members to speak up when facing morally challenging tasks, as silence is an accomplice of neutralization (Sykes & Matza, 1957).

Both individual and situational variables have a role when dealing with an ethical issue (Lopez, 2015, p. 208). Sometimes project "managers look to others and to the situation for hints about what is right and what is wrong behavior, . . . instead of adhering to their internally-held decisions of right and wrong" (Lopez, 2015, p. 212). As striving aggressively for success is encouraged by our customs, many project managers are compelled to succeed at all cost, even by sacrificing ethics, as the value of success replaces the value of virtue.

Only codes of ethics consistent with the project's culture and rightfully enforced can effectively impact ethical behavior. Project managers will more likely incorporate such ethics codes into their decisions from social pressure than regulations. "The pronouncement in a code of conduct that bribery is unacceptable is useless unless accompanied by guidelines for gift giving, payments to get goods through customs, and requests from intermediaries who are hired to ask for bribes" (Donaldson, 1996, p. 6).

Rules and regulations should act as ethical guidelines, as they can never replace people's values and beliefs, which constitute their ethical pillars. Up to a certain level, deterrence may have some impact in the enforcement of rules; nevertheless, beyond a threshold it becomes counterproductive, undermining motivation. The best way to create an ethical project culture is to live, work, and steer through values, empowering correct behaviors, and looking at "value created by the project" in addition to revenue and margin. Transparency, fairness, avoiding conflicts of interest, and open communication must be enforced, while simultaneously *Small Sins Allowed* and the *Line of Impunity* are eliminated. No personal interests, allegiances, or opportunities to act unethically should take precedence over pure project professional ethics.

Ethics is something that cannot be added at the end of the project like the icing on a cake, it needs to be embedded into the dough from the beginning and all through the end by continuous education, moral reminders, audits, and support from the parent organization.

Glossary

Agent: Representative who acts on behalf of other persons or organizations.

AIPM: Australian Institute of Project Management Standards.

Allegiance: Loyalty to a country, group, cause, or the like.

ANCSPM: Australian National Competency Standards for Project Management.

Anomie: Lack of moral standards in a society.

APM: Association for Project Management.

ASX: Australian Securities Exchange.

"Bigger is better": The belief that a corporation can profit from economies of scale by getting bigger, most of the time disregarding the associated risks for society at large.

Botnia: A finish pulp mill operating in the Uruguay River that was the center of a diplomatic dispute between Argentina and Uruguay between 2005 and 2010.

CGC: Australian Securities Exchange Corporate Governance Council.

Codes of ethics: A publicly available code that addresses a company's obligation to its stakeholders and the way it does business, including its values and vision. It provides guidance on ethical standards to the company's employees.

Collective patterns of thinking: Component of a culture that characterizes the customary way of judging employed by its members.

Confusion: Disorder resulting from a failure to behave predictably.

Conscious transgressions: The kind of evildoing that happens when individuals choose to follow an unethical path by violation of a law or a duty or moral principle.

Context: The set of facts and circumstances that surround a situation or event.

Contrasting views: Strikingly different perspectives.

Corporate governance: A complex and holistic subject related to economic efficiency and stakeholder welfare. Corporate governance can be thought of as a broad entity which includes the management processes and policies, laws, traditions and institutions affecting the authority, and accountability. It also includes relationships among stakeholders, disclosure of information, setting expectations, resource allocation, performance monitoring and control, and the moral purpose of a corporation. It should be framed by legality, legitimacy, and participation. (Lopez, 2015, p. 111).

CSR: Corporate social responsibility.

Cultural differences: Recognizes that people from different cultures have different ways of looking at things.

Cultural relativism: Infers that ethical standards vary across cultures.

Cultural universalism: Supports the idea that there should be a unique moral standard shared globally.

Culture: All the knowledge, tastes, manners, and values shared by a society.

Deviance: Deviate behavior. A state or condition markedly different from the norm.

Dishonesty: Lack of honesty; acts of lying, cheating, or stealing.

Double standards: The existence of two different sets of values, the stated and the real (which come from the unspoken culture).

Economic detachment: A shift in the meaning of morality according to what is assessed in economic and cultural terms.

Economization: Pursuit of shareholder value.

Enterprise environmental factors: Conditions not under the control of the project team, which influence, constrain, or direct the project (Project Management Institute, 2013, p. 29).

Ethical breakdowns: Conscious transgressions (the individual chooses to follow an unethical path) and unconscious transgressions (the individual does not even realize that he or she is making an inappropriate decision, falling prey to ethical fading or to other cognitive biases) (Lopez, 2015, p. 36).

Ethical decision-making models: These models recognize that factors in the environment (organizational, cultural, economic, and social) influence the recognition of a moral issue. Taking different approaches to the influence of opportunity, situational moderators, individual moderators, and significant others, the models advanced to the establishment of a moral intent to finally engage in a moral behavior (Lopez, 2015, p. 41).

Ethical dilemmas: They arise when managers face the need to make decisions that are misaligned with their own values (the organizational context collides with their personal ethical belief system) (Lopez, 2015, p. 67).

Ethical hierarchy: Successive levels of people's responses to ethical issues that require the use of different means according to the moral intensity of each ethical issue. The hierarchy starts with individual moral and values, followed by processes and policies, then ethical knowledge and awareness, and finally consultation and consensus.

Ethical issues: Problems or situations that require a choice between alternatives evaluated as ethical or unethical.

Ethical leadership: The activity of leading founded on ethics.

Ethical practices: A customary way of operation or behavior founded on ethics.

Ethical values: The set of established principles governing virtuous behavior.

Ethical virtue: The quality of doing what is ethical and avoiding doing what is unethical.

Ethics: A system of principles of right and wrong that are accepted by an individual or a social group, governing morality and acceptable conduct.

Ethics cube: A tool used to depict the complexities of ethical scenarios when more than one ethical sets of values are confronted with allegiances, interests, and with the context.

"Everybody else does it": Rationalization for a poor ethical choice.

Expectations: Belief about (or mental picture of) the future.

GAPPS: Global Alliance for Project Performance Standards.

GDRC: Global Development Research Center.

Global driver: Something that accounts for the global sense of values. It changes slowly over time (centuries or decades) and applies to wide geographic areas such as continents or blocks of nations (Lopez, 2015, p. 79).

GoPM: Governance of project management.

Governance: To pilot or steer, used in reference to the design of a system of rule. In the present day its meaning is broader, relating to concepts such as decisions, expectations, power (authority), performance, institutions, policies, processes, control (monitoring), and moral purpose. It involves three main characteristics: legality, legitimacy, and participation.

Honesty: The quality of being not disposed to cheat or defraud; not deceptive or fraudulent.

Industry driver: Something that accounts for a specific subjective sense of values. It changes as the economy, technology, and other factors evolve (Lopez, 2015, p. 79).

Institutional theory: A theory that emphasizes rational myths, isomorphism, and legitimacy. It centers on the more mysterious and changing expressions of social organization. It studies the procedures by which structures, schemes, rules, norms, and routines are in turn laid down as important rules of thumb for social behavior.

IPMA: International Project Management Association.

Latent functions: Unconscious or unintended functions that may reflect hidden purposes.

Legality: Lawfulness by virtue of conformity to a legal statute.

Legitimacy: The property of being genuine or valid, not being a fake or forgery.

Line of Impunity: The idea that certain ranks or positions in the social hierarchy entitle prerogatives or advantages, and that the power granted at those levels transcends the limits of control or law enforcement.

Local driver: Accounts for the national or local sense of values. It changes over years (possibly from one government to another) and applies to geographic areas ranging from a town to a number of countries (Lopez, 2015, p. 79).

Managerial myopia: Managers' willingness to sacrifice long-term shareholder value for short-term profits.

Manifest functions: Open, stated, conscious functions.

Mechanisms of corruption: The technical aspects of engaging in corrupt activities.

Meritocracy: A form of social system in which power goes to those with superior intellects.

Moral dilemma: State of uncertainty or perplexity, especially as requiring a moral choice between equally unfavorable options.

Moral intensity: A characteristic of the moral issue.

Moral imperative: Moral duty that is essential and urgent.

Moral neutralization: Techniques intended to nullify the moral effects of some previous action.

Moral reminder: A moral message, experience, or warning that helps people to remember something so that a mistake can be avoided.

Moral hazard: The lack of any incentive to guard against a moral risk.

Moral obligation: An obligation arising out of consideration of right and wrong.

Moral standards: A basis for moral comparison; a reference point against which other moral things can be evaluated. The moral ideal in terms of which something can be judged.

Morality: Concern with the distinction between good and evil or between right and wrong.

Motivation: The psychological feature that stimulates one to action toward a desired goal; the reason for the action; that which gives purpose and direction to behavior.

Motivation primal statement: People's desire for a decent job, a place to live, and to be part of the community.

"Never mind the rules": Regardless of the rules.

OECD: Organization for Economic Co-operation and Development.

Opportunity: A possibility due to a favorable combination of circumstances.

Orderly world: Organized world.

P2M: Project Management Association of Japan.

Paradigm: The generally accepted perspective of a particular discipline at a given time.

Paradigm shift: A move from one paradigm setting or context to another.

Participation: The act of sharing in the activities of a group.

Perception: The representation of what is perceived; a basic component in the formation of a concept. A way of conceiving something.

Permissiveness: A disposition to allow freedom of choice and behavior.

Personal interest: A personal sense of concern with and curiosity about something. A reason for wanting something.

PMBOK: The Project Management Body of Knowledge.

PMBOK®Guide: A Guide to the Project Management Body of Knowledge.

PMI: Project Management Institute.

PRINCE 2: PRojects IN Controlled Environments, version 2.

Principal: Shareholder, in the principal/agent theory.

Privileges: A special advantage, immunity, or benefit not enjoyed by all. A right reserved exclusively to a particular person or group.

Project driver: Accounts for the project's internal sense of values. It changes from project to project.

ROI: Return on investment.

"Rules of the game": Shared beliefs regarding informal or unwritten rules, attitudes, and expectations.

SAQA: South African Qualification Authority.

Scarcity-munificence: Refers to the availability of resources in an organization's environment.

SEBI: Securities and Exchange Board of India.

SGI: Sustainable Governance Indicators.

Shareholder theory: Assumes that the main purpose of an organization is to maximize shareholder return on investment.

Small Sins Allowed: A subjective mental model that establishes the level of a certain behavior above which adherence to ethical standards is expected. It can also be thought of as the ethical tolerance level that splits any dimension into two domains; above this level there are ethical standards to comply with and abide by, whereas below it there are no ethical concerns.

Social accountability: Responsibility to society or for some activity.

SOX: Sarbanes-Oxley Act.

Stakeholder: A person or organization with an interest or concern in the project activities.

Stakeholder theory: Recognizes the interests of stakeholders such as the workers, managers, suppliers, customers, and the community at large, while pursuing long-term objectives related to the creation of value for them. This theory emphasizes the importance of all parties who are affected, either directly or indirectly, by a project.

Subjective morality: An individual's belief regarding the appropriateness or incorrectness of an action, which corresponds to the concepts of moral sense, scruples, and conscience.

Transaction cost economics (TCE): An economic theory whose view of a project is considered mainly in contractual terms, assuming utility maximization self-interest, concern about shareholder control over management, and assigning preponderance to the protection of investors' capital.

Uncertainty avoidance: One of Hofstede's (1984, p. 83) value dimensions. It is defined as the degree to which the members of a society feel uncomfortable with uncertainty and ambiguity; it relates to formalization: the degree of structure in the social environment with which people feel comfortable.

Unsatisfactory equilibrium: Palliative endeavors that are unable to shape the future.

Unconscious transgressions: The kind of evildoing that happens when the individuals do not even realize that they are making an inappropriate decision to follow the unethical path by violation of a law or a duty or moral principle.

Unethical behavior: Action that does not conform to approved standards of social or professional behavior.

Unethical choice: Selection of a choice that does not conform to approved standards of social or professional behavior.

WGI: The World Governance Index.

Workplace deviance: Violation of organizational norms.

References

Alderfer, C. P. (1972). *Existence, Relatedness, and Growth: Human Needs in Organizational Settings,* New York: Free Press.

Alvarez-Dionisi, L. E. (2008). Defining the Polysemic Concept of Project Governance. In: Gutierrez, J. P., *Project Management: Methodologies and Case Studies in Construction and Engineering* (49–58). Valladolid: INSISOC Social Systems Engineering Centre, University of Valladolid.

Alvarez-Dionisi, L., and Turner, J. R. (2012) Project governance: Reviewing the past, envisioning the future. In: *Project Management Institute (PMI) Research and Education Conference 2012;* 15–18 July 2012, Limerick, Ireland.

Anderson, A. R., and Smith, R. (2007). The Moral Space in Entrepreneurship: An Exploration of Ethical Imperatives and the Moral Legitimacy of Being Enterprising. *Entrepreneurship & Regional Development*, 19(6), 479–497.

Andrews, K. R. (1971). *Concept of Corporate Strategy.* Homewood, IL: Irwin.

Ardichvili, A., Mitchell, J., and Jondle, D. (2009). Characteristics of Ethical Business Cultures. *Journal of Business Ethics,* 85(4), 445–451.

Ardichvili, A., Jondle, D., Wiley, J., Cornacchione, E., Li, J., and Thakadipuram, T. (2012). Ethical Cultures in Large Business Organizations in Brazil, Russia, India, and China. *Journal of Business Ethics*, 105(4), 415–428.

Ariely, D. (2008). *Predictably Irrational.* New York: Harper Collins.

Association for Project Management (2012). *APM Body of Knowledge (Sixth Edition).* Buckinghamshire: APM.

Australian Securities Exchange (ASE) Corporate Governance Council (2007). *Corporate Governance Principles and Recommendations with 2010 Amendments (2nd Edition).*

Bandura, A., Barbaranelli, C., Caprara, G. V., and Pastorelli, C. (1996). Mechanisms of Moral Disengagement in the Exercise of Moral Agency. *Journal of Personality and Social Psychology,* 71(2), 364–374.

Barnard, C. I. (1938). *The Functions of the Executive.* Boston: Harvard University Press.

Baucus, M. S., Norton, W. I., Jr., Baucus, D. A., and Human, S. E. (2008). Fostering Creativity and Innovation without Encouraging Unethical Behavior. *Journal of Business Ethics,* 81, 97–115.

Beenen, G., and Pinto, J. (2009). Resisting Organizational-Level Corruption: An Interview with Sherron Watkins. *Academy of Management Learning & Education,* 8(2), 275–289.

Bitektine, A. (2008). Legitimacy-Based Entry Deterrence in Inter-Population Competition. *Corporate Reputation Review,* 11(1), 73–93.

Brady, N. (1990). *Ethical Managing: Rules and Results.* New York: Macmillan.

Bredillet, C. (2014). Ethics in project management: Some Aristotelian insights. *International Journal of Managing Projects in Business,* 7(4), 548–565.

Brooks, D. (2012). Why Our Elites Stink. *The New York Times,* July 13, A23.

Burgelman, R. A., and Maidique, M. A. (1988). *Strategic Management of Technology and Innovation.* Homewood, IL: Irwin.

Canary, H. E., and Jennings, M. M. (2008). Principles and Influence in Codes of Ethics: A Centering Resonance Analysis Comparing Pre- and Post-Sarbanes-Oxley Codes of Ethics. *Journal of Business Ethics,* 80, 263–278.

Carroll, A. B. (1978). Linking Business Ethics to Behavior in Organizations. *SAM Advanced Management Journal,* 43(3), 4–11.

Chetkovich, C., & Kirp, D. L. (2001). Cases and Controversies: How Novitiates Are Trained to Be Masters of the Public Policy Universe. *Journal of Policy Analysis and Management,* 20(2), 283–314.

Cohan, W. D. (2014). *The Price of Silence: The Duke Lacrosse Scandal, the Power of the Elite, and the Corruption of Our Great Universities.* New York: Scribner.

Crawford, L. (2008). Research Paradigm of the Governance School. *Special EDEN Doctoral Seminar: The Nine Schools of Project Management.* Ecole Supérieure de Commerce Lille, France.

Cullen, J. B., Parboteeah, K. P., and Hoegl, M. (2004). Cross-National Differences in Manager Willingness to Justify Ethically Suspect Behaviors: A Test of Institutional Anomie Theory. *Academy of Management Journal,* 47(3), 411–421.

Davis, J. L, Tyge Paine, G., and McMahan, G. C. (2007). A Few Bad Apples? Scandalous Behavior of Mutual Fund Managers. *Journal of Business Ethics,* 76, 319–334.

Dayen, D. (2015). The Biggest Outrage in Atlanta's Crazy Teacher Cheating Case. *The Fiscal Times.* April 3, 2015.

De George, R. T. (1999). *Business Ethics.* Upper Saddle River, NJ: Simon & Schuster.

De Graff, F., and Herkströter, A. J. (2007). How Corporate Social Performance Is Institutionalized within the Governance Structure. *Journal of Business Ethics,* 74(2), 177–189.

Donaldson, T. (1996). Values in Tension: Ethics Away from Home. *Harvard Business Review,* 74(5), September/October.

Donaldson, T., and Dunfee, T. (1999). When Ethics Travel: The Promise and Peril of Global Business Ethics. *California Management Review,* 41(4), 43–63.

Dougherty, C. (2008). The Sheriff at Siemens Sees an Endless Battle. *The New York Times,* October 6.

Drucker, P. F. (1954). *The Practice of Management*. New York: Harper & Row.

Drucker, P. F. (1986). *Management: Tasks, Responsibilities, Practices*. New York: Truman Talley Books.

Drucker, P. F. (1993). *Post-Capitalist Society*. New York: Harper Business.

Dubbink, W., Graafland, J., and van Liedekerke, L. (2008). CSR, Transparency and the Role of Intermediate Organizations. *Journal of Business Ethics*, 82, 391–406.

Dubinsky, A. J., and Loken, B. (1989). Analyzing Ethical Decision Making in Marketing. *Journal of Business Research*, 19(2), 83–107.

Durham, G. J. (2010). Ethics: Influencing Good Choices. Retrieved from http://www.pmi.org/About-Us/Ethics/Ethics-Influencing-Good-Choices.aspx.

Durkheim, E. (1893/1964). *The Division of Labor in Society*. New York: Free Press.

Durkheim, E. (1897/1966). *Suicide: A Study in Sociology*. New York: Free Press.

Elkington, J., and Hartigan, P. (2008). *The Power of Unreasonable People*. Boston: Harvard Business Press.

Emerson, R. W. (1983). *Essays and Lectures*. New York: Library of America.

Ferrell, O. C.. and Gresham, L. G. (1985). A Contingency Framework for Understanding Ethical Decision Making in Marketing. *Journal of Marketing*, 49(3), 87–96.

Fitzgerald, F. S. (1925 (2013). *The Great Gatsby*. New York: Scribner.

Forum for a New World Governance (2012). World Governance Index (WGI) Why Should World Governance Be Evaluated, and for What Purpose? Retrieved from http://www.world-governance.org/IMG/pdf_WGI_short_version_EN_web-4.pdf.

Fraedrich, J. P. (1992). Signs and Signals of Unethical Behavior. *Business Forum*, 17(2), 13–17.

Freeman, R. E. (1984/2010). *Strategic Management: A Stakeholder Approach*. Cambridge: HarperCollins/Cambridge University Press.

Freeman, R. E., and Gilbert, D. E., Jr. (1988). *Corporate Strategy and the Search for Ethics*. Englewood Cliffs, NJ: Prentice Hall.

French, J., and Raven, B. (1959). The Bases of Social Power. In: D. Cartwright (Ed.), *Studies in Social Power*, 150–167. Ann Arbor, MI: Institute for Social Research.

Friedman, M. (1970). The Social Responsibility of Business Is to Increase Its Profits. *The New York Times Sunday Magazine* (September 13), 32–33, 122–124.

Garland, R. (2009). *Project Governance*. London: Kogan Page.

Ghemawat, P. (1991). *Commitment: The Dynamic of Strategy*. New York: Free Press.

Global Development Research Center (2015). UN-ESCAP: What Is Good Governance? Retrieved from http://www.gdrc.org/u-gov/escap-governance.htm.

Grisham, T. (2011). PMI & IPMA: Differences & Synergies. Retrieved from http://www.allpm.com/index.php/free -resources/94-article/newsletter-article/164-pmiipma.

Hamermesh, R. G. (1986). *Making Strategy Work: How Senior Managers Produce Results*. Hoboken, NJ: John Wiley.

Harford, J., Kecskes, A., and Mansi, S. (2015). Do Long-Term Investors Improve Corporate Decision Making? Retrieved from http://gsm.ucdavis.edu/sites/main/files/file-attachments/01_do_long_term_investors_improve_corporate_decision_making.pdf.

Harper, D. (1990). Spotlight Abuse—Save Profits. *Industrial Distribution*, 79, 47–51.

Harris, J., and Bromiley, P. (2007). Incentives to Cheat: The Influence of Executive Compensation and Firm Performance on Financial Misrepresentation. *Organization Science,* 18(3), 350–367.

Herzberg, F. (1966). *Work and the Nature of Man.* Cleveland: World Publishing Company.

Hillman, A., and Keim, G. (2001). Shareholder Value, Stakeholder Management, and Social Issues: What's the Bottom Line? *Strategic Management Journal,* 22, 125–139.

Hofstede, G. (1984). Cultural Dimensions in Management and Planning. *Asia Pacific Journal of Management,* 1(2), 81–99.

Hollinger, R. C., and Clark, J. P. (1983). *Theft by Employees.* Lexington: Lexington Books.

Hosmer, L. T. (1994). Strategic Planning as if Ethics Mattered. *Strategic Management Journal,* 15, 17–34.

Huehn, M. P. (2008). Unenlightened Economism: The Antecedents of Bad Corporate Governance and Ethical Decline. *Journal of Business Ethics,* 81, 823–835.

Hunt, S. D., and Vitell, S. (1986). A General Theory of Marketing Ethics. *Journal of Macromarketing,* 6(1), 5–16.

Jackson, I. A., and Nelson, J. (2004). Values-Driven Performance: Seven Strategies for Delivering Profits with Principles. *Ivey Business Journal,* November/December, 1–8.

James, B. (1994). Narrative and Organizational Control: Corporate Visionaries, Ethics and Power. *The International Journal of Human Resource Management,* 5(4), 927–951.

Jennings, M. M. (1999). *Business Ethics: Case Studies and Selected Readings (3rd Edition).* New York: West Educational Publishing.

Johnstone, D., Huff, S., and Hope, B. (2006). IT Projects: Conflict, Governance, and Systems Thinking. *Proceedings of the 39th Annual Hawaii International Conference on System Sciences.*

Jones, T. (1991). Ethical Decision Making by Individuals in Organizations: An Issue-Contingent Model. *Academy of Management Review,* 16(2), 366–395.

Joyner, B. E., and Payne, D. (2002). Evolution and Implementation: A Study of Values, Business Ethics and Corporate Social Responsibility. *Journal of Business Ethics,* 41, 297–311.

Kawabata, Y. (1969). *House of the Sleeping Beauties.* Palo Alto, CA: Quadriga.

Kelemen, M., and Peltonen, T. (2001). Ethics, Morality and the Subject: The Contribution of Zygmunt Bauman and Michel Foucault to Postmodern Business Ethics. *Scandinavian Journal of Management,* 17, 151–166.

Kennedy, C. (1993). Changing the Company Culture at Ciba-Geigy. *Long Range Planning,* 26, 18–27.

Kish-Gephart, J., Harrison, D., and Trevino, L. (2010). Bad Apples, Bad Cases, and Bad Barrels: Meta-analytic Evidence about Sources of Unethical Decisions at Work. *Journal of Applied Psychology,* 95(1), 1–31.

Klakegg, O. J., Williams, T., Magnussen, O. M., and Glasspool, H. (2008). Governance Frameworks for Public Project Development and Estimation. *Project Management Journal,* 39(1), S27–S42.

KPMG (2008). International Survey of Corporate Responsibility Reporting. Retrieved from http://www.kpmg.com/EU/en/Documents/KPMG_International_survey_ Corporate_responsibility_Survey_Reporting_2008.pdf.

Kvalnes, Ø. (2014). Honesty in Projects. *International Journal of Managing Projects in Business*, 7(4), 590–600.

Laufer, W. S., and Robertson, D. C. (1997). Corporate Ethics Initiatives as Social Control. *Journal of Business Ethics*, 16, 1029–1048.

Lawler, E. (2015). Reps. Todd Courser and Cindy Gamrat: The evolution of a sex scandal. Retrieved from: http://www.mlive.com/lansing-news/index.ssf/2015/08/ reps_todd_courser_and_cindy_ga.html.

Lee, C. H. (1990). Corporate Behavior in Theory and History: I. The Evolution of Theory. *Business History*, 32(1), 17–31.

Levine, M. (2005). *Broken Windows, Broken Business: How the Smallest Remedies Reap the Biggest Rewards*. New York: Business Plus.

Lopez, E. (2015). Influence of Ethical Behaviors in Corporate Governance. Dissertation submitted for the degree of Doctor of Philosophy in Strategy, Programme and Project Management, SKEMA Business School, Lille, France, 2 February 2015.

Lopez, E., and Medina, A. (2015). Influence of Ethical Behaviors in Corporate Governance. *International Journal of Managing Projects in Business*, 8(3), 586–611.

Lorange, P. (1980). *Corporate Planning*. Upper Saddle River, NJ: Prentice Hall.

Ludlow, P. (2013). The Banality of Systemic Evil. *The New York Times*. Retrieved from http://opinionator.blogs.nytimes.com/2013/09/15/the-banality-of-systemic-evil/?_ r=0.

Maddex, R. L. (1995). *Constitutions of the World*. New York: Routledge.

Manzoni, J. F. (2012). Building and Structuring a High Performance–High Integrity Corporate Culture. INSEAD Faculty and Research Working Paper. Retrieved from http://www.insead.edu/facultyresearch/research/doc.cfm?did=49327.

Maslow, A. (1943). A Theory of Human Motivation. *Psychological Review*, 50(4), 370–396.

Mazar, N., Amir, O., and Ariely, D. (2008). The Dishonesty of Honest People: A Theory of Self-Concept Maintenance. *Journal of Marketing Research*, 45(6), 633–644.

McCoy, C. S. (1985). *Management of Values: The Ethical Difference in Corporate Policy and Performance*. Boston: Ballinger.

Merton, R. K. (1938). Social Structure and Anomie. *American Sociological Review*, 3(5), 672–682.

Merton, R. K. (1949/1968). *Social Theory and Social Structure*. New York: Free Press.

Merton, R. K. (1968). The Matthew Effect in Science. *Science*, 159, 56–63.

Messner, S., and Rosenfeld, R. (2001). *Crime and the American Dream*. Wadsworth, CA: Cengage Learning.

Miles, R. E. (1982). *Coffin Nails and Corporate Strategies*. Englewood Cliffs, NJ: Prentice-Hall.

Milgram, S. (1975). *Obedience to Authority: An Experimental View*. New York: Harper & Row.

Mintzberg, H., and Quinn, J. B. (1991). *The Strategy Process: Concepts, Contexts, Cases*. Upper Saddle River, NJ: Prentice-Hall.

Müller, R. (2009). *Project Governance*. Surrey: Surrey Gower.

Müller, R., Andersen, E., Kvalnes, Ø., Shao, J., Sankaran, S., Turner, J. R., Biesenthal, C., Walker, D., and Gudergan, S. (2013). The Interrelationship of Governance, Trust, and Ethics in Temporary Organizations. *Project Management Journal*, 44(4), 26–44.

Müller, R., Turner, R., Andersen, E. S., Shao, J., and Kvalnes, Ø. (2014). Ethics, Trust, and Governance in Temporary Organizations. *Project Management Journal*, 45(4), 39–54.

Nace, T. (2003). *Gangs of America: The Rise of Corporate Power and the Disabling of Democracy*. San Francisco: Berrett-Koehler.

Ohmae, K. (1982). *The Mind of the Strategist: The Art of Japanese Business*. New York: McGraw-Hill.

Organization for Economic Co-operation and Development (OECD) (2004). OECD Principles of Corporate Governance. Retrieved from http://www. oecd.org/corporate/ca/corporategovernanceprinciples/31557724.pdf.

Pettigrew, A. (1992). On Studying Managerial Elites. *Strategic Management Journal*, 13.

Pinto, J., Leana, C. R., and Pil, F. K. (2008). Corrupt Organizations or Organizations of Corrupt Individuals? Two Types of Organizational-Level Corruption. *Academy of Management Review*, 33(3), 685–709.

Piper, T. R., Gentile, M. C., and Parks, S. D. (1993). *Can Ethics Be Taught? Perspectives, Challenges, and Approaches at Harvard Business School*. Boston: Harvard Business School Press.

Porter, E. (2012). The Spreading Scourge of Corporate Corruption. *The New York Times*. Retrieved from http://www.nytimes.com/ 2012/07/11/business/economy/the-spreading-scourge-of-corporate-corruption.html?_r=0.

Porter, M. (1980). *Competitive Strategy: Techniques for Analyzing Industries and Competitors*. New York: Free Press.

Porter, M. E. (1985). *Competitive Advantage: Creating and Sustaining Superior Performance*. New York: Free Press.

Porter, P. K., and Russell, G. M. (2004). Review of "Gangs of America: The Rise of Corporate Power and the Disabling of Democracy" by Ted Nace (2003). *Academy of Management Review*, 29(4), 689–691.

Prahalad, C. K., and Doz, Y. L. (1987). *The Multinational Mission: Balancing Local Demands and Global Vision*. New York: Free Press.

Project Management Institute (2012). Code of Ethics and Professional Conduct. Retrieved from http://www.pmi.org/en/About-Us/Ethics/~/media/PDF/Ethics/ap_pmicodeofethics.ashx.

Project Management Institute (1987). *Project Management Body of Knowledge (PMBOK)*. Newtown Square, PA: Project Management Institute.

Project Management Institute (1996). *A Guide to the Project Management Body of Knowledge (PMBOK Guide)*. Newtown Square, PA: Project Management Institute.

Project Management Institute (2006). *The Standard for Program Management*. Newtown Square, PA: Project Management Institute.

Project Management Institute (2008). *A Guide to the Project Management Body of Knowledge (PMBOK Guide) Fourth Edition*. Newtown Square, PA: Project Management Institute.

Project Management Institute (2013). *A Guide to the Project Management Body of Knowledge (PMBOK Guide) Fifth Edition*. Newtown Square, PA: Project Management Institute.

Pryke, S. D. (2005). Towards a Social Network Theory of Project Governance. *Construction Management and Economics, 23,* 927–939.

Quinn, J. B. (1980). *Strategies for Change: Logical Incrementalism.* Homewood, IL: Irwin.

Raiborn, C. A., and Payne, D. (1990). Corporate Codes of Conduct: A Collective Conscience and Continuum. *Journal of Business Ethics, 9,* 897–889.

Raterman, J. (2003). How to Survive the Special Project. *Journal of Banking and Financial Services* [online]. Available from: http://www.highbeam.com [Accessed: 10 Nov 2015].

Renz, P. (2007). *Project Governance: Implementing Corporate Governance and Business Ethics in Nonprofit Organizations.* New York: Physica-Verlag.

Rest, J. R. (1986). *Moral Development: Advances in Research and Theory.* New York: Praeger.

Robertson, C. J. (2008). An Analysis of 10 Years of Business Ethics Research in *Strategic Management Journal:* 1996–2005. *Journal of Business Ethics, 80,* 745–753.

Robinson, S. L., and Bennett, R. J. (1995). A Typology of Deviant Workplace Behaviors: A Multidimensional Scaling Study. *Academy of Management Journal,* 38(2), 555–572.

Ross, D. L., and Benson, J. A. (1995). Cultural Change in Ethical Redemption: A Corporate Case Study. *The Journal of Business Communication,* 32(4), 345–362.

Rumelt, R. P., Schendel, D. E., and Teece, D. J. (1994). *Fundamental Issues in Strategy: A Research Agenda.* Boston: Harvard Business School Press.

Schaefer, R. T. (2005). *Sociology.* New York: McGraw-Hill.

Schmidt, M. S., and Wyatt, E. (2012). Corporate Fraud Cases Often Spare Individuals. *The New York Times,* B1. Retrieved from http://www.nytimes.com/2012/08/08/business/more-fraud-settlements-for-companies-but-rarely-individuals.html?_r=0.

Securities and Exchange Board of India (SEBI) (2003, Feb. 8). Report of the Securities and Exchange Board of India (SEBI) Commitee on Corporate Governance. Retrieved from: http://www.sebi.gov.in/cms/sebi_data/attachdocs/1293094958536.pdf.

Selznick, P. (1957). *Leadership in Administration: A Sociological Interpretation.* Berkeley, CA: University of California Press.

SGI (2015). Sustainable Governance Indicators. Retrieved from http://www.sgi-network.org/2015.

Simon, H. A. (1945). *Administrative Behavior.* New York: Free Press.

Stanford, J. H. (2004). Curing the Ethical Malaise in Corporate America: Organizational Structure as the Antidote. *SAM Advanced Management Journal,* 69(3), 14–21.

Stark, A. (1993). What's the Matter with Business Ethics? *Harvard Business Review,* 71(3), 38–46.

Stevens, J. M., Steensma, H. K., Harrison, D. A., and Cochran, P. L. (2005). Symbolic or Substantive Document? The Influence of Ethics Codes on Financial Executives' Decisions. *Strategic Management Journal,* 26, 181–195.

Sweeney, P. (2008). Corporate Governance across Borders: Do Market Maturity and Culture Matter? *Academy of Management Perspectives,* 22, 118–119.

Sykes, G., and Matza, D. (1957). Techniques of Neutralization: A Theory of Delinquency. *American Sociological Review,* 22(6), 664–670.

Taketomi, T. (2009). Towards Establishing Project Governance. In Ohara, S., and Asada, T. (Eds.), *Japanese Project Management: KPM—Innovation, Development and Improvement.* Singapore: World Scientific.

The Economist (2008). The Siemens Scandal: Bavarian Baksheesh. Retrieved from http://www.economist.com/businessfinance /displaystory.cfm?story_id=12814642# (Dec. 17).

Thomas, L. (2006). The Winding Road to Grasso's Huge Payday. *The New York Times.* Retrieved from: http://www.nytimes.com/2006/06/25/business/yourmoney/25grasso.html?pagewanted=all&_r=0.

Thomson, L. D. (2003). Memorandum to Heads of Department Components and United States Attorneys. Retrieved from https://www.skadden.com/sites/default/files/ckeditorfiles/Thompson-Memorandum.pdf.

Thorne, L., and Saunders, S. B. (2002). The Socio-cultural Embeddedness of Individuals' Ethical Reasoning in Organizations (Cross-cultural Ethics). *Journal of Business Ethics,* 35, 1–14.

Trevino, L. K. (1986). Ethical Decision Making in Organizations: A Person-Situation Interactionist Model. *The Academy of Management Review,* 11(3), 601–617.

Tricker, R. I. (Ed.) (2000). *Corporate Governance (History of Management Thought).* Aldershot: Gower.

Turner, J. R. (2009). *The Handbook of Project-Based Management: Leading Strategic Change in Organizations.* London: McGraw-Hill.

Turner, J. R., and Müller, R. (2004). Communication and Cooperation on Projects between the Project Owner as Principal and the Project Manager as Agent. *European Management Journal,* 21(3), 327–336.

United States Mission to the Organization for Economic Co-operation and Development (2015). http://usoecd.usmission.gov/corporate_governance.html.

Van Gigch, J. P. (2008). The Ethics of Governance Projects: Ethics in Personal and Public Lives. *Systems Research and Behavioral Science,* 25, 151–156.

Washington, M., and Zajac, E. J. (2005). Status Evolution and Competition: Theory and Evidence. *Academy of Management Journal,* 48(2), 282–296.

Whetten, D. A., Rands, G., and Godfrey, P. (2007). What Are the Responsibilities of Business to Society? In Pettigrew, A., Thomas, H., and Whittington, R. (Eds.), *Handbook of Strategy and Management,* 373–408. London: Sage.

Wideman, M. (2014). The Origins of the Project Management Body of Knowledge. Retrieved from http://www.maxwideman.com/musings/origins.htm.

Woods, T. (2010). *New York Daily News.* Retrieved from http://www.nydailynews.com/entertainment/gossip/tiger-woods-admits-thought-entitled-doesn-deserve-forgiveness-article-1.197884#ixzz2Vol9pbTa.

World Bank (2006). *A Decade of Measuring the Quality of Governance.* Washington: The International Bank for Reconstruction and Development.

World Bank (2012). Importance of Corporate Governance. Retrieved from http://rru.worldbank.org/Documents/Toolkits/ corpgov_module1.pdf.

Index